Introduction to Statistics

A Computer Illustrated Text

INTRODUCTION
TO
STATISTICS

A W Bowman
University of Glasgow

and

D R Robinson
Brighton Polytechnic

Adam Hilger, Bristol

British Library Cataloguing in Publication Data

Bowman, A.W.
  Introduction to statistics: computer
  illustrated packs.
  1. Mathematical statistics
  I. Title   II. Robinson, D.R.
  519.5   QA276

ISBN 0-85274-408-0
ISBN 0-85274-409-9 Set
ISBN 0-85274-410-2 BBC 40-track disc
ISBN 0-85274-411-2 BBC 80-track disc
ISBN 0-85274-412-9 IBM PC disc

Series Editor: **R D Harding**, University of Cambridge

Published under the Adam Hilger imprint by IOP Publishing Limited
Techno House, Redcliffe Way, Bristol BS1 6NX, England

Typeset by Mathematical Composition Setters Ltd, Salisbury, UK
Printed in Great Britain by WBC Print Ltd, Bristol

⟩ Contents

〉 Preface

Statistical methods are used in a tremendously wide variety of contexts, including areas as diverse as medical research, business planning and investigations into the authorship of old texts. As a result, an increasing number of school, college and university courses require familiarity with the more commonly used statistical techniques. Proper use of these techniques requires not only knowledge of their 'mechanics'—use of formulae, tables and so on—but also an understanding of their scope and limitations. In this text + software package we have attempted to use the computer both to help describe the application of statistical techniques and to illustrate their rationale. The approach that we have adopted is to explore each topic through detailed discussion of examples. Mathematical argument is kept to a minimum, although the general theory is stated succinctly, and with verbal explanation, at some point in each section. Exercises are provided for the reader to test his or her understanding.

The software consists of a suite of ready-made programs which illustrate, and allow the reader to explore, the material covered in the text. The programs are designed to be simple to use, and no previous knowledge of computers, or of programming, is required. A principal feature of the software is that points in the text are illustrated by the use of graphics, which can be modified as the argument of the text demands. Simulations are also used, to illustrate the meaning of 'randomness' and to investigate the properties and behaviour of some of the statistical techniques. At some points, animation is also used to demonstrate clearly how some techniques are carried out in practice. One of the most important features is that you, the reader, are in charge of what you see. Demonstrations may be repeated, with different options or data sets, until the ideas involved have been consolidated and a 'feel' for the principle has been developed. We hope that the graphic displays will also be found helpful by teachers as demonstrations to illustrate and enliven lessons and lectures.

The text contains the material covered in most first courses on statistics. In Chapter 1, some of the statistician's most useful tools, such as the histogram, stem and leaf diagram and summary statistics, are introduced. Chapter 2 provides a brief introduction to those aspects of probability theory required in the rest of the text. (This subject is discussed in more depth in a companion volume: *Introduction to Probability*.) Chapters 3, 4 and 5 move on to the core of statistical inference, dealing with estimation, confidence intervals and hypothesis tests. The computer programs allow the discussion of important topics such as robustness and power, which are often not treated in elementary texts. The final chapter describes a number of the more commonly used statistical tests and the software illustrates both the theory and the practice of these. A set of statistical tables is provided in Appendix 4.

The primary purpose of the software is to help explain and illustrate ideas introduced in the text. However, several of the programs have a facility for reading data from a file on disc, and this greatly increases their flexibility as a teaching tool. The use of datafiles, together with the editing facilities provided within these programs, provides the software with a secondary role as a statistical package, capable of performing a range of simple statistical analyses on small sets of data. Of course, the facilities provided are not as extensive as those of a purpose-built package for routine analysis, because the principal aim of the program is to explain ideas.

In order to get started, you should consult Appendix 1, which describes how to load and run the software. The text may then be read with the micro to hand. Instructions on the use of the software are given in the text as you proceed. Appendix 2 describes how to create and store files of data on disc for later use. Do not feel too constrained in your use of the computer by the exercises given in the text—you will get the most out of the software if you experiment.

We are very pleased to acknowledge the assistance of a great many colleagues and students who have commented on earlier versions of the text and programs, in particular at the Statistical Laboratory, University of Manchester, and the Department of Mathematics, Statistics and Operational Research at Brighton Polytechnic. Particular thanks are also due to Robert Harding, the Series Editor, to Professor Ray Streater, the

Consultant Editor for Mathematics, to Marcus Bowman for advice on programming, and to Professor Adelchi Azzalini. The numerical routines used in the software for finding probabilities and percentage points of standard distributions are adaptations of algorithms given in *Statistical Computing*, by W J Kennedy and J E Gentle (Marcel Dekker, New York) and *Pocketbook of Mathematical Functions*, edited by M Abramowitz and I A Stegun (Verlag Harri Deutsch, Frankfurt am Main). Tables 1 to 4 were constructed with the help of NAG Fortran subroutines. Finally, our thanks go to Janet and Stephanie for their continued patience and encouragement during the preparation of this work.

〉 Chapter 1

〉 Displaying and Summarising Data

In the minds of many people, the word 'statistics' conjures up tables of numbers and the calculation from them of obscure numerical facts. In fact, the subject of statistics is concerned with establishing the story that data have to tell—and the strength of conviction with which they tell it. The context of the data may involve the applicability of a scientific law in a new situation, the relative merits of two treatments for a medical condition, the likelihood of a particular party winning the next general election or the probable sales of a product about to be launched onto the market. The analysis of such data will often involve not only the technical skills of statistics and knowledge of the background to the data, but also communications skills, in providing a clear and truthful summary of what the data do, and do not, tell us.

Actually, statistical ideas have a great deal to offer even before data are collected. For instance, they may indicate how data should be collected in order to make the best use of available resources, and they may give guidance on how many data it is necessary to collect.

In this chapter, we shall be discussing some simple methods of summarising data by the use of diagrams and numbers. Despite their simplicity, these methods can go a long way towards revealing what the data have to say, and they will also often be useful in communicating our conclusions.

〉1.1　Graphical display

Graphical display of data forms an important part of any analysis, principally because it allows us to 'explore' the data in an informal way.

We may develop and apply sophisticated techniques for analysis, but if we have not examined the data to make sure that the techniques are appropriate then we may find at a later stage that we have reached invalid conclusions and wasted our time. Common sense has a very important role to play in statistics! Another good reason for representing data graphically, or for summarising them, is that this can be a very useful way of expressing our results and conclusions when the analysis is complete. An informal examination can also serve as a check on the conclusions of more formal methods. In this chapter we will look at a variety of examples, illustrating different kinds of data, and discuss ways of representing the data graphically and ways of summarising them usefully.

Illustration 1.1

A geologist has been contracted by a mining company to survey an area where traces of a valuable mineral have been discovered. As part of his examination, the geologist takes away a number of rock samples of similar size and subjects these to laboratory analysis to determine the concentration of the mineral in each specimen. The results are given in the table below, in units of milligrams per gram.

Table 1.1   Concentration (in mg g^{-1}) of mineral in rock samples.

58	17	17	29	13
14	3	26	22	6
7	19	28	18	31
75	18	37	10	18
28	16	19	12	15
34	5	41	20	9
17	16	27	11	38
22	30	12	9	49
18	8	16	64	38
12	21			

This is an example of an experiment in which the outcome is a measurement on a *continuous* scale. This means that the results can take any value within some interval. Of course, in practice the measurements recorded will be governed by the accuracy of our measuring instrument and in the present case measurements have been made only to the nearest one milligram per gram.

There are some simple comments that we might make after examining this table. For example, all the observations are greater than 0 and less than 100. However, most people find graphical information easier to assimilate than lists of numbers. One of the most widely used graphical techniques for this kind of data is the histogram.

Histograms

In order to see an illustration of the construction of a histogram, run program STEM. (See Appendix 1 for step-by-step instructions on loading and running the programs.) When the program begins you will be given the option of which dataset to examine. The default dataset is the one of Example 1.1, called Concn, so when the prompt 'Filename?' appears on the screen simply press ⟨RETURN⟩† (i.e. press the RETURN key; do not type RETURN!). When the data have been loaded the screen will display the observations in a tabular form similar to that of table 1.1. The question we wish to discuss now is how best to display the data in graphical form.

At the bottom of the screen you will see a coloured box with the word 'Histogram' written on it, indicating the current option that the program is offering. This convention for displaying the options open to you at a particular stage is widely used throughout the book. If you press the space bar (actually, nearly any key will do but the space bar is usually the most convenient) the next option will be displayed. If you press the space bar repeatedly you will be able to 'scroll' through all the available options. In the present case there are only three options, labelled 'Histogram', 'Stem & leaf plot' and 'Transform the data' and these repeat in cyclic fashion if scrolling continues. A particular option is selected by first scrolling to it and then pressing the ⟨RETURN⟩ key.

Select the 'Histogram' option. The result is that an axis is drawn near the bottom of the screen. The scale of the axis is calculated to cover the range of the observations and the axis is marked with notches at a number of evenly spaced positions. The program has halted so that you can follow what is happening at your own pace. To continue, simply press the space bar. As you watch, the first number, 58, is moved across the screen and placed on the axis between the notches at 50 and 60. This is illustrating what we do when we construct a histogram. First of all we split the axis up into a number of 'bins' and then count the number of observations that fall into each bin. In the illustration which the program

† The RETURN key is called ENTER on the IBM PC.

is displaying, the bins are identified by the section of the axis between adjacent notches and the observations literally 'fall' into their appropriate bin! We have to decide what should happen to observations whose value coincides with the position of one of the notches. When this happens we shall assign such an observation to the bin on the right-hand side. This means that the bins refer to the ranges

[0,10), [10,20), [20,30) ... and so on

where a square bracket indicates that an end-point is included in the range and a curved bracket indicates that it is not.

Once you are clear on what is happening you will probably wish to speed up the process; this can be done simply by pressing the space bar. If the space bar is pressed repeatedly, or held down for a little while, then the animation will become very fast indeed, although the speed moves up an increment only after each number has been deposited in its appropriate bin.

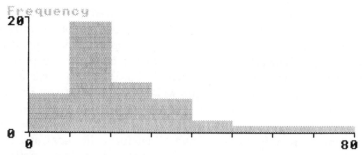

Figure 1.1   A histogram of the concentration data, produced by program STEM.

The histogram is completed by the addition of a frequency scale, so that we can see how many observations have been placed in each bin. A full and proper description of the vertical scale is that it records 'Frequency per $10\,\text{mg}\,\text{g}^{-1}$ interval'. The display that should be on your screen at this stage is reproduced in figure 1.1. We are now in a position to make use of our graphical display. In particular, we are able to assess the 'shape' of the distribution of the observations. We can see that the interval 10–20 has the largest number of observations, and that the frequency falls away very rapidly below this interval but falls away at a slower rate above it. In other words, the observations are 'stretched out' to a greater degree on the right-hand side of the picture than on the left.

This is described by saying that the distribution of the observations is *skewed* to the right. Such information is not easily identified directly from table 1.1. It can be important to identify skewness. Not only is it itself an interesting feature of the data, but it may affect the way in which we apply more formal methods of analysis, such as those to be discussed in subsequent chapters. We shall return to this point later.

Notice that when we construct a histogram we inevitably lose some of the information contained in the original data. For example, we now know by looking carefully at the picture that there are seven observations lying between 0 and 10, but we are no longer able to say exactly what these observations are without referring back to the original table. (Replace the data on the screen by pressing the space bar.) This seems a small price to pay in exchange for a picture which gives very useful information about the data. There is, however, a way of obtaining such a picture while at the same time retaining virtually all the information in the sample. This can be useful if we are drawing pictures by hand and, for example, later decide to alter the number of bins.

Stem and leaf plots
At this stage with program STEM, the option displayed at the bottom of the screen is 'Stem & leaf plot'. This is the name of a graphical display which gives a picture of the same form as a histogram but constructed from the digits of the observations themselves. Select this option by pressing ⟨RETURN⟩ and ignore for the moment the message printed near the foot of the screen. By pressing the space bar, you will see again a scale drawn on the screen covering the range 0–79. However, this time the scale is drawn vertically rather than horizontally and is marked in rather an unusual way. At each position there is a digit followed by a *. For example, consider the position 5*. This is a way of indicating that this section of the scale refers to two-digit numbers which have a 5 in the 'tens' position. So, it refers to any number in the range 50–59. A similar meaning is given to all the other sections of the scale. Numbers in the range 0–9 are considered for this purpose to be represented as 00–09. This scale, created from the first digit of the observations, is referred to as the 'stem' of the plot.

If you now press the space bar to start the animation you will see each number floating across the screen as before. However, as each one approaches the stem of the plot you will notice that the first digit of the number moves into position to identify itself with the appropriate part of the stem, while the second digit positions itself in the same row on the right-hand side of the stem. These digits are referred to as the 'leaves', whence the name 'stem and leaf' plot. As the animation continues, the leaves build up on the right-hand side of the plot. The process can be speeded up as before by pressing the space bar as often as you wish. At the end, a histogram shape has been created from the digits of the observations, although it may not be immediately recognisable by you as such. The histogram shape is now on its side, rotated through 90° in a clockwise direction. To finish the plot we can, for neatness, order the observations attached to each stem and record the frequency for each bin. The computer will do this for you if you press the space bar.

The advantage of this kind of display is that the original observations are still recoverable from it. We can read off the entire set of observations very easily as 03, 05, 06, etc. The stem and leaf plot therefore combines a useful picture with no loss of information. This sort of display is very easy to construct with paper and pencil and, when examining data without the aid of the computer, you may find it easier to construct stem and leaf plots than histograms.

The present example is a particularly convenient one for stem and leaf plots because the data consist of integers between 0 and 80. We should consider how to construct such plots when the scale of the observations is not so simple and when the bin lengths are not so convenient. As a simple example, suppose we decide that the bins for the present set of data should be 0–19, 20–39, 40–59, 60–79, 80–99. First of all we have to construct an appropriate stem. We cannot use the 'tens' digit to do this because these are not always the same within a stem, so we have to revert to the 'hundreds' digit, with the 'tens' digit forming the leaves. Press the space bar to see one possible way of labelling the stem. Notice that the message at the bottom of the screen says '2 represents 20', reminding us that the interpretation of the plot is that the pair '02' refers to the number '020', i.e. 20. The stem consists of the 'hundreds' digit followed by a '*' or a letter. The '*' refers to tens in the range 0–1; 'T' stands for two and refers to the range 2–3; 'F' stands for four and refers to the range 4–5; 'S' stands for six and refers to the range 6–7; 'E' stands for eight and refers to the range 8–9.

What we have done so far demonstrates that a stem and leaf plot can be constructed by first representing each observation as a 'stem digit' followed by a 'leaf digit'. In the first of the displays we constructed, the stem digits were the tens and the leaf digits were the units. In the second plot, the stem digits were the hundreds and the leaf digits were the tens. In each case a message was printed to remind us of the meaning of each stem–leaf pair on the original scale of measurement, so that in the first case 07 represents the number 7 whereas in the second case 07 represents 70. Notice that in the second case we have had to discard the units digit and truncate the numbers to the tens digit. For each dataset that you examine in program STEM the computer will attempt to choose a convenient stem. It will, however, also draw the plot based on a stem corresponding to a smaller number of bins. In the present example, the first plot looks preferable since it gives a little more detail in the display and also retains the information in the units digit.

The following worked examples and exercises demonstrate how stem and leaf plots may be constructed under a variety of scales of measurement.

Worked example: change of scale
Suppose that when the data of Illustration 1.1 were collected, the geologist recorded his measurements in grams instead of milligrams per gram. The data then consist of the numbers 0.058, 0.017, 0.017, etc. How would we have constructed a stem and leaf plot?

This problem can be illustrated in STEM by selecting the 'Transform the data' option. You will then be asked to enter a function of X. Since we wish to convert from milligrams to grams the appropriate function in this case is $X/1000$. X may be represented either in upper case or in lower case. When $\langle$RETURN$\rangle$ is pressed this transformation is applied to each observation and the new set of data is displayed on the screen. The 'Stem & leaf plot' option may now be selected once again.

Our first concern is to find a suitable way of representing each number in stem + leaf form. For this set of data the 'hundredths' digit might form the basis of a suitable stem, with the leaves constructed from the 'thousandths' digits. When the space bar is pressed the numbers are rewritten as two digits, with a message at the bottom of the screen to remind us that these are not to be read as integer numbers but as stems + leaves, which in fact refer to a different scale. In this example, the digits 58 represent the number 0.058, which is represented on the screen in exponential, or scientific, notation as $5.8E - 2$. (This means 5.8 multiplied by 10 to the power -2.) As we would expect, the stem and leaf plot that is constructed in this way is identical to the original one we drew, with the exception of this message about the scale of the data.

Worked example: changing the number of stem positions
As part of a market research exercise, a clothes store has persuaded a

sample of men who have purchased a particular style of garment to
supply their ages. These are given below.

29	23	19	25	21	28	52	42
19	25	23	29	26	33	37	26
37	26	24	18	27	28	21	44
22	44	32	28				

Using paper and pencil, draw a stem and leaf plot of these data, using
the tens digits to construct the stem and the units digits to form the
leaves. Now use program STEM on these data. You will first have to use
the EDITOR to create a datafile. Appendix 2 gives step-by-step instructions on how to do this. Now run program STEM and in response to the
'Filename?' prompt give the name by which you called the datafile. How
does the computer's stem and leaf plot compare with yours? You probably chose the stem to be of the form 1*, 2*, 3*, 4*, etc. If you did so
then the plot you drew would have all the data bunched into a small
number of bins. The computer took account of this by choosing a stem
of the form 1F, 2*, 2F, 3*, 3F, etc, where * refers to leaves in the range
0–4 and F, standing for five, refers to the range 5–9. In this way, the
plot has a more suitable scale. The plot with 1*, 2*, etc as the stem is
also produced by STEM when the number of bins is halved.

  We have now seen three different ways of scaling the stem of a plot.
One is to allocate leaves 0–9 to each stem position, another is to group
the leaves into the ranges 0–1, 2–3, 4–5, 6–7, 8–9 and a third is to use
the grouping 0–4, 5–9. These three possibilities are enough to give us all
the flexibility we need in choosing the stem scale of a stem and leaf plot.

Worked example: leading and trailing digits
A microcomputer manufacturing company has carried out a series of
spot checks among retailers to find the prices at which one of its models,
whose recommended price is £499.90, is actually being sold. These prices
are listed below in pounds sterling.

479.90	465.00	475.00	439.50	455.00
420.00	475.00	475.00	465.00	410.00
490.00	480.00	425.00	489.90	470.00
439.90	455.00	430.00	475.00	475.00
455.00	475.00	459.50	460.00	

Use the EDITOR again to place these numbers in a datafile, from which they may be read by program STEM. Before you ask the computer to construct a stem and leaf plot, think about how we would do this with paper and pencil.

Run program STEM with this set of data and select the 'Stem & leaf plot' option. For convenience, the leading digit, which we know is always 4, is stripped off each number and the trailing digits, representing the pence, are also removed. The meaning of the stems and leaves is recorded for us by the message that '79 represents $400 + 79$'. If we were to construct a stem and leaf plot by hand, it would probably be neater to leave the leading 4 in position and have the stem as 41∗, 42∗, 43∗, etc. However, it is easier to tell the computer to routinely strip off leading digits that are identical.

Notice that the final display with this set of data is skewed to the left. Most of the retailers are quoting prices which are just a little less than the recommended price of £499.90, but quite a few are offering quite large discounts, so that the distribution is stretched out in the direction of lower prices.

Exercise: drill and practice
The 'Transform the data' option of program STEM can be used to practice the construction of stem and leaf plots through the expression RND(1), which will generate a random number between 0 and 1. For example, with the default dataset Concn loaded, select the 'Transform the data' option and enter the expression '$20 + 10∗RND(1)$'. This will cause a set of numbers between 20 and 30 to be generated. Use paper and pencil to construct a stem and leaf plot of these and check your answer using STEM. Carry out the same exercise on data generated by the following expressions in the 'Transform the data' option:

(i)	RND(1)	(ii)	$1 + RND(1)$
(iii)	$4 + RND(1)∗10$	(iv)	$30∗RND(1)$
(v)	$1000∗RND(1)$	(vi)	$9.5 + 10∗RND(1)$

(Notice that the original display of the data may occasionally be affected by rounding.)

Exercise: negative numbers
Program STEM will not allow negative numbers to be analysed, and transformations which would lead to numbers less than zero are forbidden. This is simply a matter of programming convenience and space. Now that you have seen the principles in action with positive data,

consider how you would construct a stem and leaf plot to display the
following data, which refer to the maximum November temperatures, in
degrees centigrade, over 21 consecutive years at a certain weather station:

2.1	-1.3	0.2	2.2	-3.5	-1.6	-1.5	0.3
0.1	0.3	-0.4	-0.2	-0.8	1.1	-1.4	0.4
-1.5	3.0	1.8	0.2	-2.2	-0.4		

(Hint: a suitable stem for the positive part of these data might be 0*, 1*,
2*, 3*. Try 'reflecting' this into the negative axis, so that the stem below
0* would consist of $-0*$ for observations from -0.0 down to -0.9,
$-1*$ for observations from -1.0 down to -1.9, etc.)

It is a necessary part of the process of constructing a histogram or stem
and leaf plot to divide the axis into a number of bins so that the numbers
of observations in each of these small intervals can be recorded. This
raises the question of what is the best number of bins to use. Program
STEM attempts to choose a convenient number of bins, usually about
8, but also goes on to display the plot based on about half this number
of bins. This feature can be examined in more detail using program
HIST. Run this program and choose the default dataset, which is that
of Illustration 1.1 again. Select the first displayed option, which is
'Histogram'. This time the picture is drawn very quickly and the option
which is now on display is 'Change the number of histogram bins'. Select
this option and a message, or *prompt*, is displayed, inviting a number to
be entered. The number 8 is displayed in brackets near the end of the
prompt. This is the number of bins that is currently in use and is the
default value. If the RETURN key is pressed at this stage then the
default value will be used and the picture will remain unchanged.
However, if a different number is entered the new value will be used (and
will become the default value on the next occasion). Try entering the
number 4. After a moment's pause for calculation, the histogram is
redrawn. With such a small number of bins the histogram oversum-
marises the data and much detail is lost. Select the 'Change the number
of histogram bins' option again and this time enter the number 32. Now
there are so many bins that each one contains a very small number of
observations and there is too much detail to discern the overall shape of
the distribution of the observations. The number we started with, namely
8, was after all quite a reasonable compromise between loss of detail and
loss of shape. For any particular dataset, an appropriate number of bins
can be selected in this way simply by examining the shapes of the

histograms produced. One 'rule of thumb' which is sometimes a useful starting point is to make the number of bins approximately equal to the square root of the sample size.

The last two examples in this subsection draw attention to two features that we might look out for when we attempt to interpret the shape of a histogram.

Worked example: bimodality
An entomologist is studying the physical characteristics of a rare beetle and, as part of his investigations, he has measured the body lengths of a sample of these. These are given below in millimetres.

1.53	1.21	1.72	2.01	2.03	1.48	1.81	1.31
1.86	1.24	1.92	1.86	1.64	1.91	1.32	1.56
2.17	1.83	1.91	1.46	1.02	1.47	1.42	

Use program STEM or HIST to represent these numbers graphically and consider what information the plot is conveying. There seems to be a cluster of data at values near 1.5 mm, and another cluster near 1.9 mm. A histogram or stem and leaf plot which has this kind of feature is said to be displaying *bimodality*. A *mode* is a point at which the frequency of occurrence is higher than neighbouring points; hence the description 'bimodal' implies that we have two modes in the data. There may be several reasons why bimodality occurs. In the present case, we shall see later that the two clusters correspond to male and female beetles.

Worked example: outliers
Run program HIST again, with the default data on mineral concentrations. Suppose that the data collected by the geologist had been exactly as in Illustration 1.1 with one exception, that the last observation in the list was not 21 but 310. We can examine the effect of this by making use of the editing facility in program HIST. First select the 'Edit the data' option. You will be invited to supply the case number of the observation you wish to alter. Enter 47, followed as usual by ⟨RETURN⟩. You are now prompted for the value of this observation, with the current value (21) supplied as the default. Enter instead the value 310. In response to another prompt for a case number simply press ⟨RETURN⟩ and you may then move on to select other options with the altered set of data.

Select the 'Histogram' option and compare the shape produced with the one for the original set of data. (A reminder of what this looks like is given in figure 1.1.) The alteration of a single observation has greatly changed the picture. This has happened because the new observation is quite far from the rest of the data and it raises the question of why this observation is so unlike all the others. There are many possible reasons for this, one obvious one being that the number has actually been recorded wrongly, perhaps because an error has been made in writing it down or because the equipment that makes the measurements had a malfunction at that point. On the other hand, the value may be quite correct and may reflect the fact that a small number of samples will have a much higher mineral concentration than the others. Whatever the reason, this observation requires further investigation and, in the meantime, we may wish to remove it from the sample and consider the other observations on their own.

Observations which lie at some distance from the rest of the data are referred to as *outliers*. These observations merit special treatment, including an investigation of the causes of such unusual behaviour. Of course, it is not always easy to decide whether an observation is an outlier or not.

Line diagrams
Histograms and stem and leaf plots are invaluable aids in that they display a great deal of information about the shape of the distribution of data. However, these displays are likely to be of use only if we have a reasonably large number of observations. If we have only a small number of observations, say up to 10 or 12, then it may be better to use a different kind of display. One possibility is the line diagram. Run program TESTS2 with the default set of data. This consists of the beetle data considered above, but with the additional information of the sex of each beetle recorded (1—male, 2—female).

Table 1.2   Body lengths (in millimetres) of male and female beetles.

Males					Females				
1.24	1.31	1.42	1.56	1.72	1.02	1.21	1.32	1.46	1.47
1.81	1.83	1.86	1.91	1.92	1.48	1.53	1.64	1.86	1.91
2.01	2.03	2.17							

There are 13 male beetles and 10 female beetles. This is too few observations to draw separate histograms for each group. When the 'line plot' option is selected, two identical axes are drawn on the screen. On the top one, the body length of each male beetle is indicated by a dot. (Any character would do; an x is a useful alternative.) On the bottom axis the same thing is done for the female beetles. From this kind of display we can at least see the range covered by the data and other simple features. We are also able to make comparisons between the two groups of data, noting for instance that the males do tend to have larger body lengths than the females.

Cumulative relative frequency diagrams
The methods of display that we have so far considered have largely been concerned with the frequencies associated with values or intervals. An alternative approach which is sometimes useful is to base a diagram on the *cumulative* frequencies (or relative frequencies) up to each value. Run program HIST with the default data and select the 'Cumulative relative frequency plot' option. As you watch, you will first of all see the observations sorted into increasing order. Then a horizontal axis is drawn at the foot of the screen in the usual fashion, but this time the vertical axis refers to the *cumulative relative frequency*. For any value of x, we plot the *proportion* of observations which are less than or equal to x. (In the *cumulative frequency* plot the *number* of observations that are less than or equal to x is plotted.) The 'staircase' line on the screen displays exactly this.

Let us see why the plotted function has steps in it. If x is chosen to be small enough, then there are no observations below it and so the cumulative relative frequency at this point is zero. This remains true until x is increased to take the value of the smallest observation in the sample, namely 3. Since we now have one observation that is less than or equal to x, the cumulative relative frequency is 1/47 and so there is a step in the function at this point. In general, as x moves between any two observations the number of observations to the left of it remains unchanged. When x reaches the next observation the number of observations less than or equal to x immediately increases by one. This means that the cumulative relative frequency jumps up a step of size $1/n$ (where n is the sample size) at each data point. If some observations happen to take identical values then the step size will be a multiple of $1/n$ at this point.

This function is also referred to as the *empirical distribution function*. In mathematical terms, it is defined by

$$F_n(x) = (\text{number of observations which are} \leqslant x)/n.$$

The empirical distribution function has a variety of uses. One application is in checking its shape against some theoretical shape of a distribution to see how well the two agree. Another is in plotting two empirical distribution functions for two samples of data and measuring the distance between these to search for possible differences between the samples. However, the use which we shall explore briefly here is in the calculation of percentiles.

The *sample p-percentile* is defined to be the point below which $p\%$ of a sample lies. A cumulative relative frequency plot is a convenient tool for identifying sample percentiles. Suppose, for instance, that we are interested in the 50th percentile, which, as we shall see in § 1.2, is a useful way of identifying the 'centre' of a sample. If we move to the 0.5 position on the vertical axis of our plot and then proceed across the graph, travelling parallel to the horizontal axis, we will at some point hit the cumulative relative frequency curve. The x value of this point could be defined as the 50th percentile, since the cumulative relative frequency takes the value 0.5 here. This is, in fact, a little easier to do if we slightly alter the shape and definition of our curve. Press the space bar and the graph will be redrawn, this time with a rather smoother curve. What we have done is to interpolate between the steps of the original curve in a way that is consistent with the use of percentiles in §1.2. At this stage the program will invite you to supply a proportion to define the pth percentile. Enter 0.5. Two straight lines are drawn on the plot, illustrating the process by which we can read off the 50th percentile from the x axis. An accurate value for this is printed near the top of the screen. When you wish to withdraw from this option, simply enter the value 0, or press ⟨RETURN⟩ and scrolling will be reactivated.

Notice also that when percentiles are requested only proportions between $1/(n + 1)$ and $n/(n + 1)$ are acceptable. Outside this range there are no longer two points between which interpolation can be carried out.

Barcharts
The data we wish to examine may not always be measured on a continuous scale, as the following illustrations show.

Illustration 1.2
A small airline which specialises in short domestic flights is carrying out a review of its passenger routes. The following information refers to the number of passengers carried on daily flights in one direction between

two cities during one week of operation.

Monday Tuesday Wednesday Thursday Friday Saturday Sunday
65   42   26   54   89   102   121.

Illustration 1.3
A quality-control inspector is examining a machine which adds colour to thread and which is suspected of having a fault that occasionally causes slight imperfections in colouring. The inspector has chosen at random 50 sections of thread coloured by the machine, each 100 m long, and counted the number of imperfections in each.

Number of imperfections: 0 1   2   3 4 5 6 7 8
Frequency: 6 8 10 12 8 4 0 1 1.

The distinguishing feature of this kind of data is that the observations are no longer measurements on a continuous scale, but are essentially discrete. In the first example, the data are collected into separate groups, or categories, referring to the different days of the week. This is an example of *categorical* data. In the second case the data are indexed by the integers 0,...,8 and so the observations are measured on a meaningful, numerical scale, although not a continuous one. In this case we refer to the data as *discrete*.

Since small sets of categorical or discrete data consist only of a list of frequencies, there may seem to be less need for a graphical display. The observations are already grouped, and the 'shape' of the distribution is much more easily identified from the numerical information than in the continuous case. However, it can be helpful to reinforce this with a graphical display.

Run program FREQ and load the data of Illustration 1.2 by pressing A when the menu appears. The data are now listed along with a display at the top of the screen. This display looks very much like a histogram, representing the frequency of each group by the height of a bar, but since the groups are now separate categories and not sections of a continuous scale, narrower bars have been drawn to indicate the separation between the categories. Such a plot is referred to as a *barchart*. In this example, we can see clearly the pattern of travel throughout the week, with a rise at the weekend and a dip in the middle of the week.

Select the option 'Return to menu' and load the data of Illustration 1.3 by pressing T. This set of data is treated in a very similar way to the previous one, except that instead of having names the categories are labelled with the integer values 0–8. Although the horizontal scale is a numerical one, it is clearer not to position the vertical axis at $x = 0$. Again the 'shape' of the distribution is clear; the largest numbers are associated with thread lengths with 2 or 3 imperfections, and only a small number have 5 or more.

Scatterplots

Before leaving the topic of display, we shall look at one further example, introducing another type of data. In this example, a medical researcher is interested in whether (and, if so, how) the level of a protein changes in expectant mothers throughout pregnancy. Observations have been taken on 19 women at different stages of pregnancy.

Table 1.3   Protein levels throughout pregnancy.

Time into pregnancy (weeks)	Protein level (mg ml^{-1})
11	0.38
12	0.58
13	0.51
15	0.38
17	0.58
18	0.67
19	0.84
21	0.56
22	0.78
25	0.86
27	0.65
28	0.74
29	0.83
30	0.99
31	0.84
33	1.04
34	0.92
35	1.18
36	0.92

This is an example of *bivariate* data, where each observation consists of two measurements, in this case a time and a protein level. The protein levels could be analysed on their own by constructing a histogram or stem and leaf plot. However, we suspect that the protein levels change

throughout the course of the pregnancy and so it would be more interesting, and more informative, to examine how things change over time.
Run program SCATTER, which has this dataset, Protein, as the default.
This program displays the data in the form of a *scatterplot*, where each
observation is represented by its position on a graph. The display which
should be on your screen is reproduced in figure 1.2. This makes it much
easier to examine the relationship between protein level and time. We can
see clearly the steady increase over the weeks and can make a reasonable
guess at the *rate* of increase, something which it is much easier to do
from the scatterplot than from the list of data. We shall be looking at
data of this type in more detail in Chapter 6.

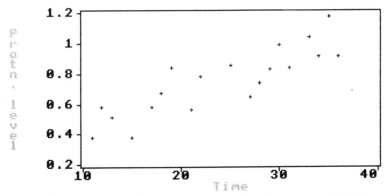

Figure 1.2   A scatterplot of the protein data, produced by program SCATTER.

We have now looked at some examples of displaying simple sets of
data graphically, and have drawn attention to the distinction between
continuous and discrete scales of measurement. We have introduced
the histogram, with its stem and leaf version, the line diagram, the
cumulative relative frequency diagram, the barchart and the scatterplot.
There are of course other ways of representing data. In general there is
no single 'best' way of displaying a particular dataset, although some
ways of displaying data are positively misleading as the book *How to Lie
with Statistics* shows (see the Bibliography). The type of display that we
choose may be governed by the particular feature of the data that we
wish to highlight or examine. One further graphical display will be introduced in § 1.2 when we discuss ways of summarising data.

In this book, you will from time to time be invited to use the computer to help you carry out some simple analyses on small sets of data. As you have seen, many of the programs have default sets of data already stored in them. Other sets of data are listed in the text and these can be typed into each program or, more efficiently, datafiles can be created as described in Appendix 2. You are, however, encouraged to collect and examine small datasets of your own. Some examples are given below.

(1) The number of cars that pass a certain point on a road in 15 different one-minute periods.
Plan this experiment before collecting the data. How should the 15 one-minute periods be chosen? What are the advantages and disadvantages of choosing them to be (*a*) successive minutes over a 15-minute period, or (*b*) 15 separate minutes, randomly spread throughout a period of one hour?
(2) The length of time it takes you to travel to school/college/work on 10 different days.
(3) The heights and weights of 20 of your friends.

In each case, how would you display the data graphically? You may like to try out some of the statistical techniques that you will be meeting in later chapters on these datasets of your own.

⟩1.2   Summarising data

In the previous section we were concerned with displaying a complete set of data with a view to examining it informally. However, there are many occasions when what we want to do is simply to summarise the data. Rather than present the whole set of numbers in tabular or even graphical form, it would be very much more convenient and economical to present one or two summary numbers. Numbers which are derived from observed data are called *statistics*. Since we are constructing these to form a summary of the data we shall refer to them as *summary statistics*. In this section we will discuss how we might do this.

Consider again the thread sections of Illustration 1.3. Run program FREQ and display the data again in barchart form. There are many ways in which we might summarise the information in this set of data. For example, we might give the smallest and largest observed values, 0 and 8 in this example. The maximum and minimum, or *extremes*, of the data tell us the range of values covered by the sample. However, the commonest way of summarising a set of data is to describe where it is centred and then to describe the scale of the spread of observations about this centre.

Sample mean and standard deviation

The most commonly used measure of the centre is the average or *sample mean*, defined by

$$\bar{x} = \sum_i x_i/n. \qquad (1.2.1)$$

Here, the sample of data is denoted by $x_1, ..., x_n$. n denotes the total number of observations in the sample, referred to as the *sample size*. (The symbol $\sum_i x_i$ indicates that all the observations x_i are added together, i.e. $\sum_i x_i = x_1 + ... + x_n$.) The sample mean averages all the data points to provide a measure of where the distribution is centred.

When the data are discrete, as in the thread data for example, they may be reduced to a set of frequencies $f_1, f_2, ...$ etc. In this case, a slightly more convenient formula for the sample mean is given by

$$\bar{x} = \sum_j f_j x_j/n.$$

f_j refers to the number of observations in the jth class or group and so $\sum_j f_j$ is just the total number of observations in the sample, namely n. Here, the index j runs through the different classes and x_j refers to the value associated with the jth class. Since each x_j occurs in the dataset exactly f_j times, the formula for the sample mean is equivalent to (1.2.1).

Run program FREQ with the thread data and select the 'Calculate the sample mean and variance' option to see how the calculations might be carried out. The column headed 'f' refers to the observed frequency of each class, with associated value 'x'. The result, 2.62, is displayed on the barchart. Notice that 2.62 is not a possible value for any particular observation in the present example since the number of imperfections in an individual thread must, of course, be a whole number. This, however, does not diminish the use of the sample mean as a measure of the centre of the sample.

The sample mean is not the only measure of the centre of a sample and we shall meet one other very shortly. However, before leaving FREQ we can also consider how to measure the scale or spread of a sample. This can be done through the *sample variance*, defined to be

$$s^2 = \sum_i (x_i - \bar{x})^2/(n - 1). \qquad (1.2.2)$$

We are interested in the extent to which the observations are spread about the centre of the sample, so we first calculate the deviation of each observation from the sample mean, namely $x_i - \bar{x}$. However, if we are interested in measuring a distance of each observation from the mean we would do better to remove the sign of these deviations. Squaring the deviations to obtain $(x_i - \bar{x})^2$ achieves this. We can now apply the idea of a sample mean to this set of numbers so that we calculate the average squared deviations of observations from the mean. However, for reasons which we shall discuss later it is more convenient to make a slight adjustment at this stage and instead of dividing by the total number of observations we subtract one before carrying out the division.

One unfortunate feature of the way in which we have done this is that the final result is not on the scale of the original data. This has happened because we squared the deviations of the observations from the mean and so obtained a measure of spread which is expressed in units that are the square of the original units. This can be remedied by taking the square root of the sample variance, producing the *sample standard deviation*

$$s = \left(\sum_i (x_i - \bar{x})^2 / (n - 1) \right)^{1/2}.$$

For discrete data, where each value x_j is observed f_j times, the sample variance is given by

$$s^2 = \sum_j f_j (x_j - \bar{x})^2 / (n - 1).$$

Press the space bar to see the calculation of the sample variance and standard deviation for the thread data in step-by-step fashion on the screen.

The formula given above for the sample variance is not always the easiest one to use for hand calculations. A little algebra shows that the sample variance is also given by the expression

$$s^2 = \frac{1}{n-1} \left(\sum_i x_i^2 - n\bar{x}^2 \right) \qquad (1.2.3)$$

and, in its frequency data form, by

$$s^2 = \frac{1}{n-1} \left(\sum_j f_j x_j^2 - n\bar{x}^2 \right).$$

These are slightly easier quantities to compute. Program FREQ also demonstrates the calculations based on this easier formula. However, when large numbers are involved, equation (1.2.3) can lead to some numerical inaccuracies.

Expressions (1.2.1), (1.2.2) and (1.2.3) all apply immediately to data on a continuous scale. Run program HIST and draw a histogram of the default data on mineral concentration. Now select the 'Sample mean and standard deviation' option and the values of the summary statistics will be indicated by arrows. These two statistics give a great deal of useful information on the data. The sample mean tells us where the data are centred and the sample standard deviation gives an indication of the spread about this centre. A rough rule of thumb, which applies when the histogram is approximately symmetric about a central peak, is that most of the observations in a sample usually lie within a distance of two standard deviations from the mean.

Worked example: comparing sets of data
Summary statistics can also be very useful in comparing sets of data. As an example we will use the beetle data introduced in § 1.1. The entire dataset is given again here, with the male and female beetles separated into groups.

Males					Females				
1.24	1.31	1.42	1.56	1.72	1.02	1.21	1.32	1.46	1.47
1.81	1.83	1.86	1.91	1.92	1.48	1.53	1.64	1.86	1.91
2.01	2.03	2.17							

If we calculate the sample mean and variance for the male and female groups separately, we find:

$$\text{Males} \quad : \quad n = 13 \qquad \sum_i x_i = 22.79 \qquad \sum_i x_i^2 = 40.9511$$

$$\text{Females:} \quad n = 10 \qquad \sum_i x_i = 14.90 \qquad \sum_i x_i^2 = 22.8680$$

which in turn yields

	Sample mean	Sample standard deviation
Males :	1.753	0.288
Females:	1.490	0.272

The male beetles have a larger sample mean body length than the female group, while the sample standard deviations are very similar. This provides a concise but informative comparison of the two groups of data. However, we shall have to wait until Chapter 4 before we are in a position to make statements about the population of beetles as a whole, and not just the sample we have observed.

Sample median and quartiles

There are alternative ways of summarising data. For instance, another possible way of measuring the centre of a set of data is to identify the value below which half of the observations fall and above which half of the observations fall. To see how this works, rerun program HIST with the default set of data and select the 'Sort the data' option. Now, instead of the observations being written simply in the order they were received, the data are ordered from smallest to largest. This in itself is a very useful thing to have done since some features of the data can now be spotted quite easily. For example, the extremes of the data, 3 and 75, can simply be read off. We usually denote a sample of observations by $x_1, ..., x_n$, where the index i simply refers to the order in which the observations were collected. When the data have been ordered we refer to the set of ordered values as the *order statistics* and denote this by $x_{(1)}, ..., x_{(n)}$. The brackets around the subscripts indicate that the index i refers to a position in the ordered sample and not simply to the ith observation collected. For example, in the concentration data x_1 is 58 because this was the first observation taken, whereas $x_{(1)}$ is 3 because this is the smallest observation of the whole sample.

Since there are 47 observations in this sample, this means that when the data are ordered there are 23 observations that lie below observation 24 and 23 observations that lie above observation 24. In this sense, observation 24 can be described as a measure of the centre of the sample. The name given to this measure is the *sample median*. In the concentration data the sample median is 18 since this is the value of the 24th observation. In general, the sample median of a set of n observations is the '$(n + 1)/2$'th observation. This makes sense if the sample size is odd because then $n + 1$ is even and $(n + 1)/2$ is a whole number. When the sample size is even we can define the sample median to be the average of observations $n/2$ and $n/2 + 1$. So, our final definition of the sample median is

$$\text{Sample median} = \begin{cases} x_{((n+1)/2)} & \text{if } n \text{ is odd} \\ (x_{(n/2)} + x_{(n/2+1)})/2 & \text{if } n \text{ is even.} \end{cases}$$

The observations in program HIST are tabulated in four columns of approximately equal length. This means that the middle observation will occur at the bottom of the second column or the top of the third one. In this case observation 24 is at the bottom of the second column.

This is a sensible measure of the centre of a set of data, but can we do something similar to measure the spread about the centre? The sample median has divided the data into two groups, a lower half and an upper half. We can extend this idea by defining the *sample lower quartile* to be the number below which one quarter of the data lie, and the *sample upper quartile* to be the number above which one quarter of the data lie. So, the quartiles, together with the median, divide the data into four groups of equal size. The quartiles are useful summary statistics because they tell us how far we must travel away from the median before we have covered half of the data in either direction, and in this sense they measure the spread of the data about the centre. The lower and upper quartiles may be defined respectively as the '$(n + 1)/4$'th observation and the '$3(n + 1)/4$'th observation. As in the case of the median, we can interpolate when $(n + 1)/4$ and $3(n + 1)/4$ are not whole numbers. For example, suppose that n takes the value 6. Then $(n + 1)/4 = 1.75$. So, the lower quartile is 0.75 of the way between observations 1 and 2, namely $x_{(1)} + 0.75(x_{(2)} - x_{(1)})$. Similarly, the upper quartile is $x_{(5)} + 0.25(x_{(6)} - x_{(5)})$. The 'Sample median and quartiles' option in program HIST can be used to display these three summary statistics.

Notice that the median and quartiles can be found through the 'Cumulative relative frequency plot' option in program HIST, discussed in § 1.1.4. This is done by entering the values 0.5, 0.25 and 0.75 in response to the prompt for proportions to define percentiles.

Worked example: beetle data
Here we shall find the sample median and quartiles for each set of beetle body lengths given in table 1.2.

For the males, $n = 13$.
Sample median $= (13 + 1)/2$ th observation (i.e. 7th)
                                         $= 1.83$.
Sample lower quartile $= (13 + 1)/4$ th observation (i.e. 3.5th)
                                         $= 1.42 + 0.5(1.56 - 1.42)$
                                         $= 1.49$.
Sample upper quartile $= 3(13 + 1)/4$ th observation (i.e. 10.5th)
                                         $= 1.92 + 0.5(2.01 - 1.92)$
                                         $= 1.965$.

For the females, $n = 10$.

Sample median	$= (10 + 1)/2$ th observation (i.e. 5.5th)
	$= 1.47 + 0.5(1.48 - 1.47)$
	$= 1.475$.
Sample lower quartile	$= (10 + 1)/4$ th observation (i.e. 2.75th)
	$= 1.21 + 0.75(1.32 - 1.21)$
	$= 1.2925$.
Sample upper quartile	$= 3(10 + 1)/4$ th observation (i.e. 8.25th)
	$= 1.64 + 0.25(1.86 - 1.64)$
	$= 1.695$.

Exercise

Find the sample median and quartiles of the following sets of data. In each case you can check your answers by using program HIST.

(*a*) 14, 32, 21, 16, 13, 15, 25, 26, 20, 17, 19
(*b*) 126, 109, 114, 67, 190, 164, 177, 198, 143, 106
(*c*) $-8, -4, 6, 7, 0, 3, 6, -1, 1$
(*d*) 1.5, 4.5, 2.9, 8.3, 6.1, 3.2, 9.5, 6.8.

We have now discussed two sets of summary statistics for a sample of data. In order to compare the relative merits of the sample mean and standard deviation on the one hand and the sample median and quartiles on the other, run program HIST with the default data. Now indicate the sample mean and standard deviation, and the sample median and quartiles, by scrolling to and selecting the appropriate options. The resulting display shows that the mean and the median are quite close together in this example, although the mean is slightly higher. This is not surprising when we remember that our sample is skewed to the right. The observations in the right-hand tail of the sample contribute their high values to the sample mean. The sample median, on the other hand, simply counts the number of observations to the right and left and so is unaffected by the exact positions of the data points above and below it. In general, we should expect the mean to have a higher value than the median on data which are skewed to the right.

We can investigate a more extreme form of this effect by modifying the data. Select the 'Edit the data' option, which is provided in several of the programs where samples of data are analysed in detail. Let us change the value of the highest observation, number 47, which currently has the value 75. Do this by entering the case number (47, followed by

⟨RETURN⟩) and then the new value. Enter a large number, say 310. Now return to the graphical display by pressing ⟨RETURN⟩ in response to the case number prompt. The entire set of data is now listed again to show you the final form of any changes you have made. Now sort the data, redraw the histogram and produce the two sets of summary statistics. One effect of the modified observation is that the histogram looks rather different; the x axis has had to be adjusted to accommodate the new, and very large, value. However, an examination of the summary statistics shows that the sample mean has increased from 22.83 to 27.83 and the standard deviation from 15.33 to 44.08. The increases have occurred because these statistics are composed of contributions from the value of each data point. The sample median (18), on the other hand, is completely unaffected because this value still splits the sample into two groups of equal size. The quartiles are also unaffected for the same reason.

We saw in § 1.1 the effect that outliers can have on a histogram. We have now seen that the sample median and quartiles are *robust* or *resistant* to such observations. This means that these summary statistics are virtually unaffected by one or two unusual observations. This is a useful property, particularly for the purposes of description, when we would like our summary to be representative of most of the data and not unduly influenced by one or two observations.

Worked example: trimmed mean
There are ways of modifying the sample mean in order to make it less susceptible to the effects of outliers. One way of doing this is to use a trimmed mean. Here we trim off a small percentage of the observations before averaging. For example, with the present set of data we would construct a 10% trimmed mean by trimming off the highest 5% and lowest 5% of the observations. This is easy to do with program HIST. First select the 'Sort the data' option and then enter the editing facility. (Notice that the 'Sort the data' option will not be available if the data have already been sorted.) There are 47 observations here, so that 5% consists of 2 observations. Delete the two most extreme cases at each end by entering the case numbers $-1, -2, -46, -47$. Pressing ⟨RETURN⟩ will complete the editing and the sample mean can again be calculated on this reduced set of data. You will find that this is very close to the original value for the concentration data and that it has not been seriously affected by the presence of the outlier at 310.

Another point worth noting when comparing the two sets of statistics is that the sample median and quartiles consist of three numbers, whereas the sample mean and standard deviation constitute only two. In what way is the extra information contained in the median and quartiles apparent? Both sets of statistics convey information about location and scale of the sample. However, by examining the positions of the quartiles with respect to the median we are in a position to identify any skewness in the sample. We have already seen from the histogram of the mineral concentrations that some slight skewness is present. This is reflected here by the fact that the upper quartile is a little further away from the median than the lower quartile. If we take into account the relative positions of the extremes then this impression is further reinforced.

Boxplots

The sample median and quartiles, together with the extremes, can be formed into a useful graphical display known as a *box and whisker plot*, or *boxplot* for short. This is illustrated by running program TESTS2 with the default beetle data and selecting the 'Boxplot' option. The sample median and quartiles for each group were calculated earlier. The values of these statistics are now represented graphically. In each case, the central box is placed so that its ends mark the positions of the lower and upper quartiles. The vertical line drawn through the box marks the position of the sample median and the ends of the long horizontal line through the box mark the extremes of the data. This is a very useful way of representing these summary statistics. It is particularly useful for giving an impression of skewness by the position of the ends of the box and line with respect to the median. It has a further use in comparing two or more samples of data, as in the present set of data. By constructing several boxplots on the same picture, each referring to the same axis, we can gain a visual impression of whether there are differences in location, scale or skewness among the different samples. With the beetle data, the picture re-expresses the fact that the males in our sample have larger body lengths than the females, but the spreads of the samples are very similar. We may be tempted to say that there is some evidence of skewness in the male group. However, we should be very cautious in doing so when we remember that the boxplots are based on very small samples of data. In fact, all our remarks on this set of data should be tempered by the fact that the sample size of each group is very small, and so many of the conclusions that we draw will be rather tentative ones.

⟩1.3 Summary

In this chapter we have met several methods of displaying and summarising data. Methods of display include:

histograms	which give good illustrations of the shape of the distribution of a set of observations
stem and leaf plots	which give a display very similar to the histogram but retain nearly all the information in the data and are particularly easy to construct using only paper and pencil
line diagrams	which are most suited to very small sets of data
cumulative relative frequency plots	which are particularly helpful in finding percentiles
boxplots	which are a useful way of representing the summary statistics of sample median, quartiles and extremes, especially when comparing two or more samples
scatterplots	which represent data where each observation consists of two measurements.

Several other types of graphical display, and variants of the ones we have discussed here, are described in the books listed in the Bibliography.

The summary statistics that we have discussed include the following.

Sample mean and variance: these are defined by

$$\bar{x} = \sum_i x_i/n$$

$$s^2 = \sum_i (x_i - \bar{x})^2/(n-1)$$

or, for data in frequency form, by

$$\bar{x} = \sum_j f_j x_j/n$$

$$s^2 = \sum_j f_j(x_j - \bar{x})^2/(n-1).$$

These are the most widely used summary statistics and they possess many convenient properties, some of which will be discussed in Chapter 3.

Sample median and quartiles: these are defined by

Sample lower quartile $= (n + 1)/4$ th observation
Sample median $= (n + 1)/2$ th observation
Sample upper quartile $= 3(n + 1)/4$ th observation

using interpolation where necessary. These statistics can give an indication of skewness and they are less susceptible to the effects of outliers.

Features of datasets that have been highlighted by these displays and statistics include:

skewness (right or left)
bimodality
outliers.

In addition, the main conventions of the programs have been introduced, namely:

scrolling of displayed options by pressing the space bar and selection of an option by pressing the ⟨RETURN⟩ key;
entering of numbers in reply to prompts, with a default value in brackets;
the use of case numbers to provide a simple editing facility;
the use of previously created datafiles.

⟩ Chapter 2

⟩ Probability and Sampling

⟩2.1 The need for probability theory

The limitation of the type of statistics that we have discussed so far is that it confines itself to the particular data involved. Typically, though, the object of scientific or medical experimentation, or surveys carried out by businesses or government, is to draw *general* conclusions. In this context, the data actually collected are called the *sample* of observations and the totality of all observations is called the *population*. The objective of *statistical inference* is to draw conclusions about the population on the basis of a sample.

For example, suppose that a new drug is given to 20 patients with a certain medical condition and 10 subsequently recover. Interest lies in the extent to which this clinical trial yields information about the effectiveness of the drug in general. Assuming that there was 'nothing special' about the patients in the trial, we would presumably estimate the recovery rate for the new drug to be 50%. But since the trial was small, how strong are the grounds for preferring the drug to the traditional treatment, which is known to have a recovery rate of 40%? In any case, how can we be sure that there is 'nothing special' about the patients in the trial, since all patients are special in some way?

There are two essential ingredients for successful statistical inference:

(1) samples should be randomly selected from the population of interest, as far as is practicable;

(2) probability theory should be applied to describe the variability of the observations so that the degree of uncertainty in making inferences can be measured.

We shall discuss random sampling in § 2.6, but first we need to introduce some of the elements of probability theory. It should be emphasised that this is going to be a cursory introduction to the topic, mentioning only those ideas that are required later in this book. The reader with an interest in, or a need to know more about, probability theory is referred to the companion volume *Introduction to Probability* and the other texts mentioned in the Bibliography.

⟩2.2 Probability and long-run frequencies

The underlying situation we envisage when we want to apply probability theory is the *random experiment*. (Often the word 'random' is dropped when the context is clear.) This is an activity which can, in principle at least, be repeated indefinitely and whose outcomes will vary over the course of the repetitions in a random manner. The tossing of a coin and the recording of 'heads' or 'tails' or the throwing of a die and the recording of the score are two examples of random experiments. An *event* associated with a random experiment is a collection of possible outcomes. For instance, in throwing a die, one event is 'the score is 6', while another is 'the score is even'. In this case, if we threw the die and observed a score of 2, the first event has not occurred whilst the second event has occurred.

Although we may not be able to predict with certainty whether or not a particular event is going to occur, we may be able to say that some events are more likely than others. The *probability of an event* is a measure of how likely the event is to occur in a single experiment. A probability of zero corresponds to an impossible event, whilst a probability of one corresponds to an event which is certain to occur. Events which sometimes occur and sometimes do not have probabilities strictly between 0 and 1. Roughly speaking, the probability of an event is the long-run proportion of times the event would occur if the experiment were repeated a very large number of times.

Worked example
For a particular type of tablet being produced by a pharmaceutical company, the nominal weight of the active ingredient is 50 mg, but, owing to variations in the manufacturing process, the true weight varies randomly. One event of interest is 'the amount of active ingredient in a tablet exceeds 52 mg' : we shall call this event A. Suppose that we are

told that the probability of event A, denoted by $P(A)$, is 0.01. What implication does this statement have for the number of tablets with more than 52 mg of active ingredient in a sample of 100 000?

Answer. Since the large-sample proportion of times that this event will occur is 0.01, in a sample of 100 000 tablets we would expect about $0.01 \times 100\,000 = 1000$ to have more than 52 mg of active ingredient.

Probabilities obey certain rules. The *complements rule* says that if A is an event and A^c is the complementary event (i.e. the event that only occurs whenever A does not occur) then the probability of A^c, $P(A^c)$, is given by

$$P(A^c) = 1 - P(A). \qquad (2.2.1)$$

The *addition rule* relates to mutually exclusive events. Two events, A and B say, are *mutually exclusive* if they cannot occur simultaneously (i.e. there is no outcome which is in both events). In this case, the probability that either event A occurs or event B occurs, denoted by $P(A \cup B)$, is given by

$$P(A \cup B) = P(A) + P(B). \qquad (2.2.2)$$

Worked example (continued)
Suppose that another event of interest is B = 'the weight of active ingredient in a tablet is less than 48 mg'. It is known that $P(B) = 0.02$. Describe the event $A \cup B$ in words and say how the event 'the weight of a tablet is less than or equal to 52 mg' can be represented symbolically. Find the probabilities of these two events.

Answer. $A \cup B$ is the event 'either the weight of active ingredient exceeds 52 mg or is less than 48 mg' and its probability is

$$P(A \cup B) = 0.01 + 0.02 = 0.03.$$

(Note that events A and B cannot occur simultaneously.) Further, the event A^c is 'the weight of active ingredient is less than or equal to 52 mg' and its probability is

$$P(A^c) = 1 - P(A) = 0.99.$$

Exercise
The maximum daily demand for electricity from a small power station has a probability of 0.6 of exceeding 200 kW and a probability of 0.3 of exceeding 250 kW. Let A be the event that the maximum demand exceeds

200 kW and B be the event that the maximum demand exceeds 250 kW. Describe the events A^c and $(A^c) \cup B$ in words and find their probabilities.

⟩2.3   Random variables

It frequently happens that a number is associated with the outcome of a random experiment. In throwing a die, for example, the score is usually of interest; in a health survey, several numbers associated with each subject might be noted, e.g. number of brothers and sisters, height, blood pressure. When the outcome of a random experiment takes numerical values it is called a *random variable* and is often denoted by a letter near the end of the alphabet when mathematical notation is required. Thus, in the health survey, we might denote the number of brothers and sisters by X, the height by Y and the blood pressure by Z.

If the possible values of a random variable can be listed then it is said to be a *discrete random variable*. If X is discrete, its possible values are usually denoted in general by $x_1, x_2, ..., x_n$ (or $x_1, x_2, x_3, ...$ if there is an infinite number of possible values). '$X = x_i$' then denotes the event that X takes the value x_i in the experiment. Thus the probability that X takes the value x_i is written '$P(X = x_i)$' and is the long-run *relative frequency* of the value x_i (i.e. the long-run proportion of times that this value would occur in a series of the experiments). Similarly, the expressions $P(X \neq x_i)$ and $P(X < x_i)$ refer to the probabilities that the value of X is not x_i and that the value of X is less than x_i. The list of possible values of X, along with their probabilities, is called the *probability distribution* of X.

By the rules of probability, the probabilities of the discrete random variable X taking values $x_1, x_2, ..., x_n$ must satisfy

$$0 \leqslant P(X = x_i) \leqslant 1 \qquad \text{for } i = 1, 2, ..., n \qquad (2.3.1)$$

$$P(X = x_1) + P(X = x_2) + ... + P(X = x_n) = 1. \qquad (2.3.2)$$

Worked example
A die has been weighted so that the probability of a 6 is 0.2 whilst the other scores are all equally likely; find the probability distribution of the score.

Answer. We shall denote the score by X. The possible values of X are 1, 2, ..., 6. Since $P(X = 6) = 0.2$ and

$$P(X = 1) = P(X = 2) = ... = P(X = 5)$$

we can use rule (2.3.2) as follows:

$$1 = P(X = 1) + P(X = 2) + \ldots + P(X = 5) + P(X = 6)$$
$$= 5P(X = 1) + 0.2.$$

We can therefore deduce that $P(X = 1) = 0.16$, so that the probability distribution of X is

$$P(X = k) = \begin{cases} 0.16 & k = 1, 2, 3, 4 \text{ or } 5 \\ 0.2 & k = 6 \\ 0 & k \neq 1, 2, 3, 4, 5 \text{ or } 6. \end{cases}$$

When a random variable is a measurement taking values on a continuous scale it is called a *continuous random variable*. Examples are heights, weights, blood pressures, capacities, duration times etc. For instance, suppose that X is the height of a randomly chosen adult male subject in metres. Then $P(X < 1.7)$, $P(1.7 < X < 1.72)$ and $P(X > 1.8)$ correspond to the probabilities that X is less than 1.7 m, between 1.7 and 1.72 m and greater than 1.8 m, and all of these take values strictly between 0 and 1. Similarly, $P(X = 1.7)$ denotes the probability that the height of a randomly selected man is exactly equal to 1.7, but, unlike the previous cases, the probability of this happening is zero. To see that this is so, consider the sequence of probabilities $P(1.69 < X < 1.71)$, $P(1.699 < X < 1.701)$, $P(1.6999 < X < 1.7001)$, etc. It is clear that these probabilities can become as close to zero as we wish, but it is also the case that each is greater than $P(X = 1.7)$, so the last probability must be zero. A similar argument shows that $P(X = x)$ is zero for all values of x.

The probability distribution of a continuous random variable cannot, therefore, be described by specifying $P(X = x)$ for various values of x. Instead, the probability density function $f(x)$ is used. We shall, however, defer discussion of this until we have given some examples of continuous probability distributions in § 2.5.

An important relationship that may exist between two or more random variables is that of *independence*. We shall confine ourselves here to a discussion of the practical implications of this relationship and mention mathematical details in § 2.7. Two random variables X and Y are independent if knowledge of the value taken by one of them does not affect the distribution of the other. Thus, if X were the height of the first subject in a survey and Y were the height of the second, then X and Y would be independent, assuming that the subjects were selected at random. That is to say, the distribution of Y would be the same before and after the value of X had been ascertained. On the other hand, if X

and Y were the height and weight of the first subject, they would not be independent since, for instance, the value of X would affect the distribution of Y (e.g. if X was larger than average, we would expect Y to be larger than average).

In practice, random variables may be identified as being independent by noting that they are associated with independent parts of the experiment.

Worked example

X and Y are the scores on the first and second throws of a die. Are X and Y independent?

Answer. This depends on how the 'experiment' was conducted. If for each throw we were careful to shake the die thoroughly, then X and Y could be assumed to be independent. If, however, we only lifted the die a short way above the table before letting it go with a minimum amount of spin, then X and Y would not be independent—knowledge of the value of X would help in predicting the value of Y. (Notice that the answer does not depend on whether the die is biased.)

The assumption of independence is often critical in statistical inference. Whether the assumption is valid is often associated with, among other things, the method of sampling, which will be discussed further in § 2.6.

We have seen that probability distributions are used to describe the behaviour of randomly varying numbers. Frequently we are not interested in the full probability distribution but merely with some summary measures associated with it. We shall concentrate on three of these measures: the population mean, the population variance and the population standard deviation, or, more briefly, the *mean*, *variance* and *standard deviation*.

For a discrete random variable X, taking the values $x_1, x_2, ..., x_n$, the mean is defined as

$$\sum_{i=1}^{n} x_i P(X = x_i) = x_1 P(X = x_1) + x_2 P(X = x_2) + ... + x_n P(X = x_n).$$

$$(2.3.3)$$

(The definition for continuous random variables is given in § 2.5.) It can be denoted by μ, the Greek letter 'mu', for mean, or $E(X)$, the *expected value of X* or the *expectation of X*. The mean is a measure of the location of a distribution, giving an indication of where it is centred.

The link between the mean and the sample mean is as follows. If the random experiment were repeated indefinitely and the sample means of all values of X so far obtained were evaluated after each experiment, then the values of the sample means would tend to μ. Notice that the mean is a constant, whereas the sample mean is a random variable—if you take two random samples, the values of the sample means will nearly always differ.

Worked example

A microcomputer salesman has the following sales record per customer. He has probability 0.2 of selling model G, for which his commission is £30, and probability 0.1 of selling model H, for which his commission is £50. The only possible outcomes are that he sells one of model G, one of model H or makes no sale. Find his mean commission per customer.

Answer. Let $X =$ commission for a random customer, in £'s. Then $P(X = 0) = 0.7$, $P(X = 30) = 0.2$ and $P(X = 50) = 0.1$. Hence

$$\mu(\text{or } E(X)) = (0 \times 0.7) + (30 \times 0.2) + (50 \times 0.1) = 11.$$

(Notice that although the mean is also referred to as the expected value, the value £11 is not 'expected' in the sense that it is likely to occur. In fact, the value £11 *cannot* occur in a single sale.)

The (population) variance of the discrete random variable X taking the values $x_1, x_2, ..., x_n$ is defined as

$$\sum_{i=1}^{n} (x_i - \mu)^2 P(X = x_i) = (x_1 - \mu)^2 P(X = x_1)$$

$$+ (x_2 - \mu)^2 P(X = x_2) + ... + (x_n - \mu)^2 P(X = x_n). \qquad (2.3.4)$$

(Again, the definition for continuous random variables is given in § 2.5.) The variance of X is denoted by var(X) or σ^2 ('σ' being the Greek letter 'sigma'). It follows from definition (2.3.4) that the variance cannot be negative. The variance is related to the sample variance in the same 'long-run' sense that the mean and sample mean are related. Like the sample variance, the variance is a measure of spread. For instance, a small value of σ^2 would indicate that there is a high probability that X is close to μ, i.e. X is not very variable.

The *standard deviation* σ is the square root of the variance. It is a more useful measure of spread than the variance because it is in the same units as the original observations (rather than in 'squared units'). Theoretical results are, however, usually easier to express in terms of variances.

Exercise

Find the standard deviation of the commission per customer of the microcomputer salesman in the worked example.

Exercise

Evaluate the mean and variance of the score on a fair die. Obtain a fair die and throw it 50 times, recording the score for each throw. Are the sample mean and variance approximately equal to the population values?

⟩2.4   Some discrete distributions

We have seen that a discrete (probability) distribution consists of a set of possible values, x_1, x_2, ..., x_n, along with the associated probabilities, $P(X = x_1)$, $P(X = x_2)$, ..., $P(X = x_n)$. Subject only to rules (2.3.1) and (2.3.2) being satisfied, any two sets of numbers may arise. It turns out, however, that in a great many situations a distribution belonging to one of a small number of well studied families of distributions is appropriate. We shall consider here just two of these families: the binomial and the Poisson.

Binomial distributions

A binomial distribution typically arises when we are interested in the number of members of a randomly selected group of individuals that possess a certain characteristic. Some possible examples are:

(1) the number of people replying that they favour a certain political party in a survey of 1000 people;

(2) the number of defective parts found when a batch of 200 parts is inspected;

(3) the number of female offspring in a litter of size eight for a certain species of mammal;

(4) the number of heads obtained in 10 tosses of a coin.

In each case there are three critical conditions which must hold if a binomial distribution is to be obtained: (*a*) each sampled individual must have the same probability of possessing the characteristic; (*b*) the possession of the characteristic by an individual must be independent of the possession of the characteristic by any other individual (i.e. knowledge of which other individuals possess the characteristic is of no help in predicting whether a particular individual possesses the characteristic), and (c) the number of sampled individuals must be fixed in advance.

In general, suppose that there is a random sample of n individuals, that the probability that each possesses the characteristic is p and that the random variable of interest is X, the number of individuals in the sample possessing the characteristic. X is then said to have the binomial distribution with parameters n and p. This is expressed in mathematical notation as $X \sim B(n,p)$. The possible values of X are $0, 1, 2, ..., n$, and it can be shown that the probabilities are given by

$$P(X = k) = \binom{n}{k} p^k (1 - p)^{n - k} \qquad k = 0, 1, 2, ..., n. \quad (2.4.1)$$

In this formula, $\binom{n}{k}$ stands for 'n combinatorial k', which is defined as

$$\binom{n}{k} = \frac{n!}{k! \, (n - k)!} \qquad (2.4.2)$$

where $n!$ stands for 'n factorial' and is defined by

$$n! = \begin{cases} n \times (n - 1) \times (n - 2) \times ... \times 2 \times 1 & n > 1 \\ 1 & n = 0 \text{ or } 1. \end{cases} \quad (2.4.3)$$

(Factorials and combinatorials have the following interpretations. $n!$ is the number of ways of ordering n objects; for example, the three letters A, B, C can be ordered in $3! = 6$ ways (ABC, ACB, BAC, BCA, CAB, CBA). $\binom{n}{k}$ is the number of ways of selecting k objects from n objects without regard to the order of selection; for example, two letters can be selected from the letters A, B, C in $\binom{3}{2} = 3$ ways (AB, AC, BC).)

Worked Example
Evaluate (a) $\binom{5}{2}$, (b) $\binom{2}{2}$.
   Answer.
   (a)
$$\binom{5}{2} = \frac{5!}{2! \, 3!}$$

and $5! = 5 \times 4 \times 3 \times 2 \times 1 = 120$, $2! = 2 \times 1 = 2$, $3! = 3 \times 2 \times 1 = 6$; hence

$$\binom{5}{2} = \frac{120}{2 \times 6} = 10.$$

   (b)
$$\binom{2}{2} = \frac{2!}{2! \, 0!}$$

and $2! = 2 \times 1 = 2$, $0! = 1$; hence

$$\binom{2}{2} = \frac{2}{2 \times 1} = 1.$$

The combinatorial $\binom{n}{k}$ is only defined when n and k are non-negative integers and $k \leqslant n$, and its value is always a positive whole number.

Returning to formula (2.4.1), it can be shown that these probabilities do satisfy rules (2.3.1) and (2.3.2), as long as n is a positive integer and $0 \leqslant p \leqslant 1$. Using (2.3.3) and (2.3.4), the mean and variance can be shown to be

$$\mu = np \qquad \sigma^2 = np(1 - p). \tag{2.4.4}$$

Worked example

In a particular city, 20% of residents oppose a new housing scheme. A television reporter interviews three randomly chosen residents. Find the distribution of X, the number of interviewees opposing the scheme, and the mean and variance of the distribution.

Answer. Since the interviewees are chosen at random (and assuming that the reporter does not slant his questions so as to bias the replies!) X has the $B(3, 0.2)$ distribution. From formula (2.4.1),

$$P(X = 0) = \binom{3}{0}\ 0.2^0 0.8^3 = 1 \times 1 \times 0.512 = 0.512$$
$$P(X = 1) = \binom{3}{1}\ 0.2^1 0.8^2 = 3 \times 0.2 \times 0.64 = 0.384$$
$$P(X = 2) = \binom{3}{2}\ 0.2^2 0.8^1 = 3 \times 0.04 \times 0.8 = 0.096$$
$$P(X = 3) = \binom{3}{3}\ 0.2^3 0.8^0 = 1 \times 0.008 \times 1 = 0.008.$$

From (2.4.4), the mean is equal to $3 \times 0.2 = 0.6$, and the variance is equal to $3 \times 0.2 \times (1 - 0.2) = 0.48$.

Program B&P can be used to evaluate binomial probabilities and draw diagrams of binomial distributions. Run this program, select the binomial distribution and set $n = 3$, $p = 0.2$, as in the worked example. You will see the probabilities indicated by a barchart. Their values can be displayed by selecting the 'List the probabilities' option.

It is of interest to compare the shapes of binomial distributions with different values of n and p. For example, reset n and p to 10 and 0.5 respectively. Notice that the distribution is symmetric about the most probable value, $X = 5$. If you set $n = 10$ and $p = 0.3$, the resulting distribution is skewed to the right. Compare this with the $B(10, 0.7)$ distribution—they are mirror images. Now try larger values of n. You will find that the distribution is approximately symmetric for a progressively wider range of values of p about 0.5.

Exercise
It is found that a particular drug is successful in reducing the blood pressure of patients suffering from hypertension in 60% of cases. Five patients are suffering from hypertension in a certain ward and are to be treated with the drug. Use formula (2.4.1) to evaluate the probability that more than three patients have their blood pressure reduced. Check your answer using program B&P. What assumptions are you making in obtaining this answer?

Exercise
In an engine assembly factory, 2% of engines reaching the final quality check have to be sent back for retuning. Use program B&P to find the probability that out of 100 engines, more than four will need to be sent back.

Poisson distributions
A Poisson distribution typically arises when we are interested in the number of randomly occurring events observed in a fixed interval. (We use the word 'event' here in the sense of 'a happening of interest', rather than in the technical sense defined in § 2.2.) Some possible examples are

(1) the number of accidents occurring in a factory on a given day;
(2) the number of calls arriving at a telephone exchange in a 30 second period;
(3) the number of bacteria in a small sample drawn from a well mixed solution;
(4) the number of minor imperfections found in a 1 km length of glass fibre.

In each case, the necessary conditions for the occurrence of a Poisson distribution are (*a*) the events of interest must occur independently of one another, and (*b*) the probability of an event occurring in an interval depends only on the length of the interval and not its position (i.e. there is no predictable tendency for certain times/places to be preferred over others).

In general, if Y has the Poisson distribution with parameter λ (the Greek letter 'lambda'), the possible values of Y are $0, 1, 2, \ldots$ and the probabilities are given by

$$P(Y = k) = \frac{e^{-\lambda}\lambda^k}{k!} \qquad k = 0, 1, 2 \ldots . \qquad (2.4.5)$$

Here, e is the constant $2.71828\ldots$ and $k!$ is 'k factorial', defined by (2.4.3). Although, in principle, Y can take any one of an infinite number of values, for k sufficiently large $P(Y = k)$ is extremely small. It can be shown that the probability rules (2.3.1) and (2.3.2) are satisfied by these probabilities and that the mean and variance are given by

$$\mu = \lambda \qquad \sigma^2 = \lambda. \qquad (2.4.6)$$

The value of the parameter λ is often deduced from the fact that it is the mean of the distribution.

Worked example

In a certain factory, the average number of accidents per day is 1.5. Assuming that accidents occur independently of one another and that there is no predictable tendency for more accidents to occur at certain times, deduce the distribution of the number of accidents per day and find its mean and variance. What is the probability that there will be more than two accidents in a day?

Answer. Since accidents occur at random, Y, the number of accidents occurring in one day, may reasonably be assumed to follow a Poisson distribution. We are told that the mean is 1.5, so λ takes this value. From (2.4.6) the variance is also equal to 1.5. We can evaluate the probabilities of Y taking the values 0, 1 and 2 as follows:

$$P(Y = 0) = e^{-1.5}1.5^0/0! = (0.2231 \times 1)/1 = 0.2231$$
$$P(Y = 1) = e^{-1.5}1.5^1/1! = (0.2231 \times 1.5)/1 = 0.3347$$
$$P(Y = 2) = e^{-1.5}1.5^2/2! = (0.2231 \times 2.25)/2 = 0.2510.$$

Using the addition rule (2.2.2), $P(Y \leqslant 2) = 0.2231 + 0.3347 + 0.2510 = 0.8088$. The probability of more than two accidents per day, $P(Y > 2)$, is given by the complements rule as $1 - P(Y \leqslant 2) = 0.1912$.

Poisson probabilities may be evaluated using program B&P. Run the program, select the Poisson distribution and set $\lambda = 1.5$. A barchart of the Poisson distribution is displayed. The distribution is skewed to the right, as is the case for all Poisson distributions. The probabilities evaluated in the above worked example may be confirmed by selecting the 'List the probabilities' option. Compare Poisson distributions with different values of λ.

Now select the binomial distribution with $n = 50$ and $p = 0.1$, followed by the 'Poisson approximation' option. A barchart of the Poisson distribution with $\lambda = 5.0$ ($= 50 \times 0.1$) is superimposed on the barchart of the $B(50, 0.1)$ distribution. Notice that the two distributions are very similar. (λ has been chosen equal to np so that the means of the two distributions will be equal.) You can confirm that the probabilities are similar by selecting the 'List the probabilities' option. Compare the $B(70, 0.05)$ and Poisson, $\lambda = 3.5$, distributions in the same way.

What we have observed is that if n is large (greater than 50, say) and p is small (less than 0.1, say), X has the $B(n, p)$ distribution and Y has the Poisson distribution with $\lambda = np$, then

$$P(X = k) \simeq P(Y = k) \qquad k = 0, 1, ..., n$$

i.e.   (2.4.7)

$$\binom{n}{k} p^k (1 - p)^{n-k} \simeq e^{-np} (np)^k / k!.$$

(Approximation (2.4.7) can be established mathematically.) This 'Poisson approximation to the binomial distribution' can be useful since Poisson probabilities are, in general, easier to evaluate and, in more advanced work, the Poisson distribution can be easier to handle.

Worked example
Suppose that 3% of the adult population read a certain newspaper regularly. If 100 adults are selected at random, what is the probability that exactly six will be regular readers of the newspaper?

Answer. X, the number of regular readers of the newspaper in the sample, has the $B(100, 0.03)$ distribution. We could now use formula (2.4.1), but evaluation of the expression $\binom{100}{6} 0.03^6 0.97^{94}$ is somewhat troublesome. Instead we approximate by the Poisson distribution with $\lambda = 100 \times 0.03 = 3$, so that

$$P(X = 6) \simeq e^{-3} 3^6 / 6! \simeq 0.0504.$$

Exercise
Heart attack cases arrive at a hospital at an average rate of four per day. Find the probability that less than three cases arrive during one day. (Use formula (2.4.5) first and confirm your answer by using program B&P.)

Exercise
A high-energy physics experiment involves observing collisions of two types of particle. In 8% of the collisions, a certain type of fundamental particle can be detected. If 60 collisions are observed, use the Poisson approximation to find the probability that the fundamental particle is detected in exactly five experiments.

⟩2.5 Some continuous distributions

We noted in § 2.3 that if X is a continuous random variable then $P(X = x) = 0$ for each value of x, so that the probability distribution needs to be described in a different way from discrete distributions. It turns out that the most convenient description is the *probability density function* $f(x)$. Any function defined for all values of x can be a probability density function (or pdf for short) as long as (i) it is always non-negative and (ii) the area between it and the x axis is equal to one.

Suppose that the pdf of X is the function illustrated in figure 2.1(*a*). Then probabilities associated with X can be evaluated by measuring areas under the pdf. For instance, $P(1 < X < 2)$ is indicated in figure 2.1(*b*) and $P(X < 1.5)$ is indicated in figure 2.1(*c*). Notice that the condition that the area under the pdf equals one ensures that $P(-\infty < X < \infty) = 1$. Notice also that $P(1 < X < 2) = P(1 \leqslant X \leqslant 2)$ and $P(X < 1.5) = P(X \leqslant 1.5)$, since $P(X = 1) = P(X = 2) = P(X = 1.5) = 0$.

As was the case with discrete distributions, certain families of distributions turn out to be particularly useful. The most useful are the normal distributions, but we shall start by considering a rather simpler family, the uniform distributions.

Uniform distributions
A continuous random variable X has the uniform distribution between a and b if its pdf is given by

$$f(x) = \begin{cases} \dfrac{1}{b-a} & a \leqslant x \leqslant b \\[2mm] 0 & x < a \quad \text{or} \quad x > b. \end{cases} \tag{2.5.1}$$

We shall denote this distribution by $U(a, b)$. The parameters a and b can take any values, subject to $a < b$. You can see this distribution illustrated by running program DISTN and choosing the 'Uniform distribution' op-

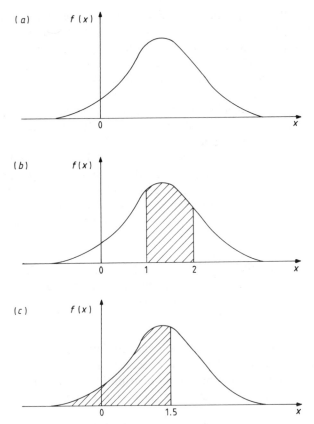

Figure 2.1   Probability density function. (*a*) pdf of *X*. (*b*) $P(1 < X < 2)$. (*c*) $P(X < 1.5)$.

tion. For instance, set *a* to 0 and *b* to 2. The shape of the pdf indicates that *X* must take values between 0 and 2 and that, roughly speaking, any value within that range is as likely as any other value. From (2.5.1), the height of the pdf between 0 and 2 is 0.5.

We can evaluate probabilities, i.e. areas under the pdf, easily for this distribution. Select the 'Evaluate $P(c < X < d)$' option and set *c* to 0.2 and *d* to 1.0, for instance. You will see that the appropriate area is shaded and the probability given as 0.4. Check this answer by evaluating the area yourself.

Another type of problem that we shall be coming across later is the evaluation of the '100*p*% point' of a distribution. This involves finding

x such that $P(X \leqslant x) = p$, where p is a given value. Select the 'Find $100p\%$ point' option and set p equal to 0.9. A left-hand area of 0.9 is shaded and the right-hand boundary is found to be 1.8, i.e. the value of x such that $P(X \leqslant x) = 0.9$ is equal to 1.8. Again, think how you could have calculated the value of the 90% point without the use of the computer.

A related type of problem involves finding the central interval of a distribution with a given probability. Select the 'Find a middle $100p\%$ region' option and set p equal to 0.9. You will see that values of c and d are found such that the area between c and d is equal to 0.9. In fact the values of c and d are chosen so that the left-hand and right-hand un-shaded areas under the pdf are each equal to 0.05. (Recall that the total area under the pdf is one.) It therefore follows that $P(X \leqslant c) = 0.05$ and $P(X \leqslant d) = 0.95$, i.e. c and d are the 5% and 95% points respectively.

Exercise
The diameters of ball bearings (in mm) have the $U(5.485, 5.515)$ distribution. Find (i) the probability that the diameter of a ball bearing lies between 5.51 and 5.52 mm and (ii) a range of diameters within which 80% of all the ball bearings fall.

You will see that the mean and standard deviation (sd) of the uniform distribution are shown at the top of the screen. Integration has to be used to find these quantities for continuous distributions. In general, if the pdf of the distribution is $f(x)$, then the mean and variance are defined as

$$\mu = E(X) = \int_{-\infty}^{\infty} x f(x) \, dx$$

$$\sigma^2 = \mathrm{var}(X) = \int_{-\infty}^{\infty} (x - \mu)^2 f(x) \, dx.$$

Normal distributions
A random variable X has the normal distribution with mean μ and variance σ^2, denoted by $N(\mu, \sigma^2)$, if its pdf is

$$f(x) = \frac{1}{(2\pi\sigma^2)^{1/2}} \exp\left(-\frac{(x-\mu)^2}{2\sigma^2}\right) \qquad -\infty < x < \infty. \quad (2.5.2)$$

It is worthwhile becoming familiar with the shape of the pdfs for various values of the parameters μ and σ^2.

Run program DISTN (or select the 'Clear the screen' option if it is already running), select the 'Normal distribution' option and set $\mu = 50$ and $\sigma^2 = 100$. You will see that the pdf is shaped like a bell and is symmetric about the mean value. Most of the area under the curve lies between about 30 and 70, though the pdf is nowhere actually equal to zero. Select the 'Normal distribution' option again and set $\mu = 60$ and $\sigma^2 = 100$, i.e. increase the mean by 10 and leave the variance unchanged. You will see that the second pdf has exactly the same shape as the first but that it is shifted 10 units to the right. Examine the shape of the $N(45,100)$ distribution also.

Now let us investigate the effect of changing the variance parameter. Select the 'Clear the screen' option and redraw the $N(50,100)$ distribution. Draw the $N(50,200)$ distribution. Note that the effect of increasing the variance is to stretch the distribution, though it retains the bell shape. Draw the $N(50,50)$ distribution. Another bell shape is obtained, but this time the distribution is more peaked than the $N(50,100)$ distribution. What do you anticipate will be the shapes of the $N(40,200)$ and $N(55,50)$ distributions? Check your answers by using program DISTN.

Because of the shapes of the pdfs, the normal family of distributions is often used when a histogram of the data approximates a bell-shaped curve. A great many sets of observations arising in practice, such as heights, blood pressures and errors in mass production processes, can be modelled by normal distributions. We shall see in § 2.7 and in later chapters, however, that there are other reasons for the importance of this family.

No simple formulae exist for probabilities associated with normal distributions, so tables have to be used. These always refer to the $N(0,1)$ distribution (the so-called *standard normal distribution*), though we shall see in a moment that they can also be used for arbitrary normal distributions. It is conventional to denote a random variable with the $N(0,1)$ distribution by Z and $P(Z \leqslant z)$ by $\Phi(z)$.

In Appendix 4, table 1, $\Phi(z)$ is tabulated against z for values of z between 0 and 3.59. In order to evaluate $\Phi(z)$ for negative values of z, use the relationship

$$\Phi(-z) = 1 - \Phi(z). \tag{2.5.3}$$

Run program DISTN and draw the $N(0,1)$ distribution. You will see that the pdf is symmetric about 0. Suppose that we wish to evaluate $P(Z < 1)$, i.e. the area under the $N(0,1)$ pdf to the left of 1. Use table 1 to verify that $\Phi(1) = 0.8413$. Now select the 'Evaluate $P(c < X < d)$' option and set $c = -10$ and $d = 1$. (In principle, c should be set to $-\infty$, but the computer would not understand this.) You will see that the appropriate area is shaded and the probability is found to be approximately 0.8413.

Worked example
Use table 1 to evaluate (i) $P(Z \leqslant -0.5)$, (ii) $P(Z > 1.5)$ and (iii) $P(0.2 \leqslant Z \leqslant 1.1)$.
   Answer. (i) From (2.5.3),

$$P(Z \leqslant -0.5) = \Phi(-0.5) = 1 - \Phi(0.5).$$

From table 1, $\Phi(0.5) = 0.6915$, so that $P(Z \leqslant -0.5) = 0.3085$.
   (ii) From the complements rule (2.2.1),

$$P(Z > 1.5) = 1 - P(Z \leqslant 1.5) = 1 - \Phi(1.5).$$

From table 1, $\Phi(1.5) = 0.9332$, so that $P(Z > 1.5) = 0.0668$
   (iii) Evaluate $P(0.2 \leqslant Z \leqslant 1.1)$ by using program DISTN. By considering the shaded and left-hand unshaded areas, note that

$$P(0.2 \leqslant Z \leqslant 1.1) = P(Z \leqslant 1.1) - P(Z \leqslant 0.2).$$

Hence

$$P(0.2 \leqslant Z \leqslant 1.1) = \Phi(1.1) - \Phi(0.2) = 0.8643 - 0.5793 = 0.285.$$

Exercise
Use table 1 to evaluate (i) $P(Z \leqslant 1.84)$, (ii) $P(Z \geqslant 0.08)$, (iii) $P(Z \leqslant -1.1)$ and (iv) $P(1.1 \leqslant Z \leqslant 1.84)$. Check your answers by using program DISTN.

   Suppose now that X has the $N(\mu, \sigma^2)$ distribution and we wish to evaluate $P(c < X < d)$ for given values of c and d ($c < d$). Our observation that all normal pdfs have a similar shape has the following implication:

$$\frac{X - \mu}{\sigma} \text{ has the } N(0,1) \text{ distribution} \qquad (2.5.4)$$

i.e. we can set $(X - \mu)/\sigma$ equal to Z and evaluate probabilities using table 1. For instance, suppose that we want to evaluate $P(X \leqslant d)$. Now if $X \leqslant d$, it follows that $X - \mu \leqslant d - \mu$ and that $(X - \mu)/\sigma \leqslant (d - \mu)/\sigma$, i.e. $Z \leqslant (d - \mu)/\sigma$. Thus

$$P(X \leqslant d) = P(Z \leqslant (d - \mu)/\sigma) = \Phi((d - \mu)/\sigma). \qquad (2.5.5)$$

Also,

$$P(X > d) = 1 - P(X \leqslant d) = 1 - \Phi((d - \mu)/\sigma) \qquad (2.5.6)$$

and

$$P(c < X < d) = P(X < d) - P(X \leqslant c) = \Phi((d - \mu)/\sigma) - \Phi((c - \mu)/\sigma). \qquad (2.5.7)$$

We now consider the evaluation of the $100p\%$ point for a given p $(0 < p < 1)$. Again, we start by considering the $N(0,1)$ distribution. We denote the $100p\%$ point of the $N(0,1)$ distribution by z_p, i.e. z_p satisfies $P(Z \leqslant z_p) = p$. In table 2 of Appendix 4, z_p is tabulated against p for selected values of p.

Worked example
Find z such that $P(Z \leqslant z) = 0.95$.
   Answer. We require the 95% point, $z_{0.95}$. From table 2, $z_{0.95} = 1.6449$. Run program DISTN, draw the $N(0,1)$ distribution and select the 'Find $100p\%$ point' option, with $p = 0.95$, to confirm this value.

Exercise
Use table 2 to find the 90% and 99% points of the $N(0,1)$ distribution. How would you use table 2 to find the 5% point? (Remember that the $N(0,1)$ pdf is symmetric about 0. Solve the problem by using program DISTN to give yourself a further hint.)

   Suppose now that X has the $N(\mu, \sigma^2)$ distribution and we wish to find the $100p\%$ point, x_p. Since $P(X < x_p) = p$,

$$P((X - \mu)/\sigma < (x_p - \mu)/\sigma) = p$$

and it follows from (2.5.4) that

$$P(Z < (x_p - \mu)/\sigma) = p$$

i.e. $(x_p - \mu)/\sigma$ is the $100p\%$ point of the $N(0,1)$ distribution. Thus

$$(x_p - \mu)/\sigma = z_p$$

i.e.

$$x_p = \mu + \sigma z_p. \qquad (2.5.8)$$

Worked example
The speeds, in mph, of different vehicles on a main road have the

$N(50,100)$ distribution. Find (i) the proportion of vehicles having speeds between 30 and 75 mph, (ii) the speed exceeded by 20% of vehicles.

Answer. (i) Let X be the speed, in mph, of a randomly selected vehicle. Then, from (2.5.7), the required probability is

$$P(30 < X < 75) = \Phi((75-50)/10) - \Phi((30-50)/10)$$
$$= \Phi(2.5) - \Phi(-2) = 0.9938 - 0.0228 = 0.9710.$$

(ii) We require $x_{0.8}$. From (2.5.8), $x_{0.8} = 50 + 10z_{0.8}$. From table 2, $z_{0.8} = 0.8416$, so that $x_{0.8} = 58.4$, i.e. the required speed is about 58 mph.

We finally consider the problem of finding a middle $100p\%$ region for a normal distribution. Use program DISTN to draw the middle 90% region of the $N(0,1)$ distribution. Notice that since the distribution is symmetric, the end points of the region are equidistant from zero. Also notice that the two unshaded areas must each be equal to 0.05 in order that the total area under the curve is equal to one. It follows that the right-hand point is just the 95% point of the $N(0,1)$ distribution, i.e. $z_{0.95}$, and can be found by using table 2.

In general, the endpoints of the middle $100p\%$ region of the $N(0,1)$ distribution are

$$- z_{(1+p)/2} \quad \text{and} \quad z_{(1+p)/2}. \tag{2.5.9}$$

If X has the $N(\mu, \sigma^2)$ distribution, then the end points of the middle $100p\%$ region for $Z = (X - \mu)/\sigma$ are also given by (2.5.9), so that the end points for the middle $100p\%$ region for X are

$$\mu - \sigma z_{(1+p)/2} \quad \text{and} \quad \mu + \sigma z_{(1+p)/2}. \tag{2.5.10}$$

Worked example
A manufacturer of ready-to-wear menswear wants his products to fit 95% of all men. If men's heights (in cm) have the $N(175,40)$ distribution, what range of heights should he cater for?

Answer. Let X be the height of a randomly selected customer, so that $X \sim N(175,40)$. We wish to use formula (2.5.10), so we must first evaluate $z_{(1+0.95)/2} = z_{0.975} = 1.96$, from table 2. The required range of heights is

$$175 - \sqrt{40} \times 1.96 \text{ to } 175 + \sqrt{40} \times 1.96$$

i.e. 162.6–187.4 cm.

Exercise
The error of a police speed measuring device, in mph, has the

$N(-0.3, 0.64)$ distribution. Within what bounds will the errors in 99.9% of cases lie?

Exponential distribution
Although the normal distributions are very important, they cannot be used to model every dataset. In particular, it often happens that the histogram of the data is positively skewed. A particularly simple distribution which might be appropriate in this case is the exponential with parameter v, which has the pdf

$$f(x) = \begin{cases} ve^{-vx} & x \geqslant 0 \\ 0 & x < 0. \end{cases} \qquad (2.5.11)$$

It can be shown that the mean and variance of this distribution are

$$\mu = 1/v \qquad \sigma^2 = 1/v^2. \qquad (2.5.12)$$

You can draw this pdf, with $v = 10$, by selecting the 'Gamma distribution' option of program DISTN and setting $k = 1$ and $v = 10$. (The exponential is a special case of the gamma distribution in which the parameter k is equal to 1.) Compare this with the pdfs of the exponential distributions with $v = 5$ and $v = 20$.

Probabilities associated with the exponential distributions are particularly simple to evaluate. If X has the exponential distribution with parameter v, then

$$P(X \leqslant x) = 1 - e^{-vx} \qquad x \geqslant 0 \qquad (2.5.13)$$

and the $100p\%$ point is

$$x_p = -\ln(1 - p)/v \qquad (2.5.14)$$

where $\ln(1 - p)$ denotes the natural logarithm of $1 - p$.

It can be shown that the random processes which give rise to Poisson distributions, when interest lies in the number of events in a fixed interval, also give rise to exponential distributions, when interest lies in the time or distance between neighbouring events (or until the first event from the start of the process). The parameter v of the exponential distribution is equal to the mean rate of the randomly occurring events.

Worked example
Recall the factory accident example, page 40, in which accidents occur at random at a mean rate of 1.5 per day. What is the probability that,

after an accident has occurred, more than two days will pass before the next accident occurs?

Answer. Let X be the time until the next accident occurs. Then X has the exponential distribution with $v = 1.5$. The required probability is

$$P(X > 2) = 1 - P(X \leqslant 2) = e^{-1.5 \times 2} = 0.05.$$

Exercise

Minor imperfections occur in the manufacture of glass fibre at an average rate of 0.4 per kilometre. Find the probability that there is no minor imperfection in the first 1 km of fibre.

Other distributions

There are a number of other commonly occurring families of probability distributions, some of which we shall encounter in later chapters. For the moment, we shall just consider briefly the gamma distributions which, as we have already seen, include the exponential distributions as a special case.

The gamma distribution with parameters k (>0) and v (>0) has the pdf

$$f(x) = \begin{cases} v^k x^{k-1} e^{-vx}/(k-1)! & x \geqslant 0 \\ 0 & x < 0 \end{cases} \qquad (2.5.15)$$

when k is an integer. You can examine the shapes of various gamma distributions by using program DISTN. You will see that they are all skewed to the right, although, for fixed v, the skewness decreases as k increases. A gamma distribution can therefore sometimes be used to model skewed data. In particular, the sum of k independent random variables, each with the exponential distribution, parameter v, has the gamma k, v distribution.

⟩2.6 Random sampling

So far we have only considered situations in which a single random variable is available. Typically, though, our data will consist of many observations which we regard as the observed values of a set of random variables. In order to be useful for making inferences, it is usually necessary to assume that the random variables are independent of one another. (This assumption is not always made, however; *time series analysis*, for instance, deals with observations which are not independent.) The assumption of independence will itself only be reasonable if the sampled units from which the observations were obtained were randomly selected.

Let us consider how such sampling can be achieved in practice, assuming, for the moment, that the population is finite. Run program SAMPLE, and select the 'Subjective sampling' option. A typical display is shown in figure 2.2. You will see illustrated up to 100 line segments, with cross pieces at both ends of each for emphasis. Enter the numbers of eight of the line segments which you judge are as representative as possible of all the line segments, in respect of the distribution of lengths. (The number of each line segment is to its left. Press RETURN after entering each number.) After having entered the eighth number, you will be told the mean and standard deviation of the lengths of the line segments in your sample, along with the population mean and standard deviation. Write down these four numbers and repeat the option a few more times.

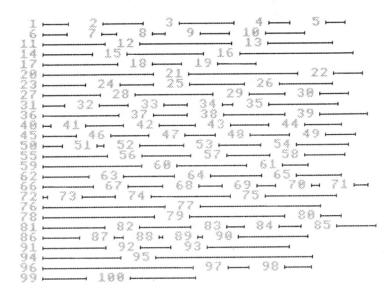

Figure 2.2 A typical display for program SAMPLE after selection of the 'Subjective sampling' option.

How well did you do? If your sample means regularly exceeded the population means, your eye was being caught too often by the longer lines. If your sample standard deviations were too high, you probably tried to avoid bias by taking equal numbers of long lines and short lines. Even if you tried to 'cheat' by taking the first line segment in each of eight rows, the chances are that your sample means were too high. (Can you see why?)

However well you succeeded, you probably now appreciate what a difficult job it is to select a 'representative sample', even in a situation where the quantity of interest is right before your eyes. In order to avoid the possibility of individual biases, it is necessary to adopt some form of random sampling procedure. We shall confine ourselves to the simplest method, *simple random sampling*.

Let us suppose that the population consists of n units and we wish to sample k of them. The units are first labelled $1, 2, 3, ..., n$. Then k different numbers are sampled at random from the set $\{1, 2, 3, ..., n\}$ (perhaps by means of drawing lots) and the corresponding units are then the ones to be included in the sample. In fact, rather than drawing lots, it is more usual to use random number tables or a random number generator.

Program SAMPLE has a 'Random number generator' option. Select it and set n to 100 and k to 8. You will see that eight different numbers from the set $\{1, 2, 3, ..., 100\}$ are chosen. If you were to repeat this option a large number of times, you would find that no particular number was favoured over any other number, nor any pair of numbers favoured over any other pair.

You can use this random number generator to select the line segments. Choose the 'Random sampling' option. Up to 100 line segments will again be illustrated, but this time the computer will select a sample of eight using the random number generator. As before, the sample and population mean and standard deviation are then displayed. Repeat this option several times and see whether, on average, the computer's samples appear to be more representative than yours.

When you do not have access to a random number generator, you will have to make use of random number tables, such as table 8 in Appendix 4. To see how these are used, select the 'Random number tables' option of program SAMPLE. You will see the first six lines of a random number table, set up in the same way as table 8, although the entries are different. In fact, in both cases each digit in the table is selected at random from the set $\{0, 1, 2, ..., 9\}$, independently of the selection of other digits in the table. (The method involved need not concern us.) The computer will now demonstrate how the table can be used to select different numbers at random from the set $\{1, 2, 3, ..., n\}$, where n is the entered population size. For instance, set n, the size of the population, to 60. You will see that the digits are considered in pairs and that the number formed by a pair is entered into the list of selected numbers if it lies between 1 and n and has not been selected previously.

A similar procedure is adopted for any value of n between 10 and 99, inclusive. For $n < 10$, single digits are considered, for $100 \leqslant n \leqslant 999$, triplets of digits are considered, and so on. Experiment with a selection of values of n. (Note that the maximum value of n accepted by program SAMPLE is 9999.)

Exercise
Use table 8 to select six numbers at random from the sets (i) $\{1, 2, 3, ..., 24\}$ and (ii) $\{1, 2, 3, ..., 500\}$.

It is often not possible to label the members of a population, perhaps because insufficient information is available about the population or because the population is too large, or, indeed, infinite. An example of the former case would be the population of all car thieves in a city, of interest to a criminologist, and an example of the latter case would be the population of all the bacteria *streptococci*, of interest to a bacteriologist. In such cases it will usually not be possible to obtain a true random sample, but an effort should be made to avoid *bias* and *non-independence*. Biased selection occurs when some members of the population are more likely to be selected than others, while non-independence occurs when the probability of a unit being sampled depends on which other units are in the sample. Biased sampling may result in observations which are systematically higher or lower than the population average, whilst non-independent sampling often leads to the sampled observations being less variable than are the observations in the whole population. The danger in the latter case is that results based on the sample may seem more precise than they should. (It is worth noting that when, as earlier in this section, sampling is without replacement, it is not true to say that the sampled observations are selected independently of one another. For instance, when selecting at random two different numbers from the set $\{1, 2, ..., 100\}$, the probability that the second number chosen is 1 is either 0 or 1/99, depending on whether or not the first number sampled was 1. This form of non-independence is not, however, a source of 'over-precise' results and, as far as making inferences is concerned, has negligible effect when the sample size is less than about one tenth of the population size.)

Worked example

It is required to select 20 students at random from a college. Two methods of sampling have been proposed: (i) choose a class at random and select, at random, 20 students from the class; (ii) ask the head of the college to select one student from each of 20 classes. In what ways are these two methods likely to be severely non-random?

Answer. (i) This sample is non-independent, since if one member of a class is selected, the other members of the class have a greatly enhanced chance of selection. The characteristics of the members of the sample will not be as varied as the characteristics of all students in the college. (ii) This sample is likely to be biased. Perhaps the head will select those students who excel in academic studies or sporting activities.

Whenever non-random sampling has to be used, it is important that the method of sampling is reported carefully, so that a judgement can be made about the wider applicability of any conclusions based on the sample.

⟩2.7   Sampling distributions

We shall suppose now that the data that we have available can be considered to be the observed values of the independent random variables $X_1, X_2, ..., X_n$, each of which has the same distribution, i.e. $X_1, X_2, ..., X_n$ form a random sample. In order to make inferences on the basis of $X_1, X_2, ..., X_n$ we shall need to know something of their joint distribution and/or the distribution of some interesting functions of them, such as the sample mean.

When the distribution of the X_i is discrete, their joint probability distribution consists of the values of $P(X_1 = x_1, X_2 = x_2, ..., X_n = x_n)$ for all possible outcomes $(x_1, x_2, ..., x_n)$. Here, $P(X_1 = x_1, X_2 = x_2, ..., X_n = x_n)$ is to be read as 'the probability that X_1 equals x_1 *and* X_2 equals x_2 *and* ... *and* X_n equals x_n', and it can be evaluated using the relation

$$P(X_1 = x_1, X_2 = x_2, ..., X_n = x_n) = P(X_1 = x_1)P(X_2 = x_2)...P(X_n = x_n).$$
(2.7.1)

Result (2.7.1), which holds if, and only if, $X_1, X_2, ..., X_n$ are independent random variables, is sometimes referred to as the *multiplication rule*. This rule allows us to calculate the distribution of functions of $X_1, X_2, ..., X_n$.

Worked example

Two fair dice are thrown; X_1 is the score of the first die, X_2 is the score of the second die. Find the joint distribution of X_1 and X_2 and the distribution of the total score, $X_1 + X_2$.

Answer. Since the dice are fair, $P(X_1 = i) = P(X_2 = i) = 1/6$ for $i = 1, 2, ..., 6$. From (2.7.1), the joint distribution is given by

$$P(X_1 = i, X_2 = j) = (1/6) \times (1/6) = 1/36 \qquad i = 1, 2, ..., 6; \; j = 1, 2, ..., 6.$$

The possible values of the total score, $X_1 + X_2$, are $2, 3, ..., 12$. The total score 2 can only occur if $X_1 = 1$ and $X_2 = 1$, so that

$$P(X_1 + X_2 = 2) = P(X_1 = 1, X_2 = 1) = 1/36.$$

The outcome $X_1 + X_2 = 3$ occurs if either $X_1 = 1$ and $X_2 = 2$ or $X_1 = 2$ and $X_2 = 1$. Thus, by the addition rule (2.2.2),

$$P(X_1 + X_2 = 3) = P(X_1 = 1, X_2 = 2) \\ + P(X_1 = 2, X_2 = 1) = 1/36 + 1/36 = 1/18.$$

The rest of the distribution of $X_1 + X_2$ can be derived in a similar manner.

The equivalent definition to (2.7.1) for continuous random variables is that the joint density function is given by

$$f(x_1, x_2, ..., x_n) = f(x_1)f(x_2) ... f(x_n). \tag{2.7.2}$$

From a knowledge of the joint distribution of $X_1, X_2, ..., X_n$, it is possible, in principle, to deduce the distribution of any function, such as $X_1 + X_2 + ... + X_n$ or $X_1 - X_2$. The method used is on the same lines as that indicated in the above worked example when finding the distribution of $X_1 + X_2$, but is, in general, much more difficult to apply.

Some simple results do, however, exist if interest is confined to the mean and variance of a function. In particular, for the linear function $\Sigma a_i X_i$, it can be shown that

$$E\left(\sum_{i=1}^{n} a_i X_i\right) = \sum_{i=1}^{n} a_i E(X_i) \tag{2.7.3}$$

$$\text{var}\left(\sum_{i=1}^{n} a_i X_i\right) = \sum_{i=1}^{n} a_i^2 \, \text{var}(X_i). \tag{2.7.4}$$

Result (2.7.4) holds if $X_1, ..., X_n$ are independent, whilst result (2.7.3) always holds. For example, if $X_1, ..., X_n$ are a random sample from a distribution with mean μ and variance σ^2, then $E(X_1 + X_2 + ... + X_n) = n\mu$ and $\text{var}(X_1 + X_2 + ... + X_n) = n\sigma^2$. Also, if $\bar{X}$ is the sample mean $(= (X_1 + X_2 + ... + X_n)/n)$ then

$$E(\bar{X}) = \mu \qquad\qquad (2.7.5)$$

$$\text{var}(\bar{X}) = \sigma^2/n \qquad\qquad (2.7.6)$$

i.e. the mean of the distribution of the sample mean is μ and the variance of the distribution is σ^2/n. This implies that if we were to take larger and larger random samples, the distribution of the sample means would always have the same mean, μ, but the variances would steadily become smaller. For sufficiently large samples, the sample mean would not be a very variable quantity and would nearly always be fairly close to the mean.

Another result of interest concerns the sample variance:

$$S^2 = \sum_{i=1}^{n} (X_i - \bar{X})^2/(n-1).$$

It can be shown that, if $X_1, ..., X_n$ is a random sample from a distribution with mean μ and variance σ^2,

$$E(S^2) = \sigma^2. \qquad\qquad (2.7.7)$$

Not so many results exist, in general, when we are interested in the full distribution of a function, as opposed to just its mean and variance. One exception is when the random sample comes from a normal distribution. For instance, in this case the distribution of the sample mean is given by

$$\bar{X} \sim N(\mu, \sigma^2/n). \qquad\qquad (2.7.8)$$

Also, if X has the $N(\mu_1, \sigma_1^2)$ distribution, Y has the $N(\mu_2, \sigma_2^2)$ distribution and X and Y are independent, then the distributions of their sum and difference are given by

$$X + Y \sim N(\mu_1 + \mu_2, \sigma_1^2 + \sigma_2^2)$$

$$\qquad\qquad (2.7.9)$$

$$X - Y \sim N(\mu_1 - \mu_2, \sigma_1^2 + \sigma_2^2).$$

Worked example

Two sisters, Kate and Sally, are keen marathon runners. Kate's times (in minutes) follow the $N(180,40)$ distribution, whilst Sally tends to be well behind her sister, her times being independent of Kate's and having the $N(260,60)$ distribution. Find (i) the probability that Kate's average time in 4 races is less than 175 min, and (ii) the probability that Sally finishes no more than 70 min behind Kate in a race.

Answer. (i) Let X_1, X_2, X_3, X_4 be Kate's four times. Then if $\bar{X}$ is the sample mean time, from (2.7.8), $\bar{X} \sim N(180,40/4)$, i.e. $N(180,10)$. From (2.5.5), the required probability is

$$P(\bar{X} < 175) = \Phi((175-180)/\sqrt{10})$$

$$= \Phi(-1.58) = 0.057.$$

(ii) Let X be Kate's time and Y be Sally's time in the race. Then, from (2.7.9), the difference in times is $Y - X \sim N(260-180, 60+40)$, i.e. $N(80,100)$. The required probability is

$$P(Y - X < 70) = \Phi((70-80)/\sqrt{100}) = 0.159.$$

The importance of the normal family of distributions rests on a remarkable result called the *central limit theorem*. Loosely speaking, this states that if X_1, X_2, ..., X_n are a random sample from a distribution with mean μ and variance σ^2 (not necessarily a normal distribution), then if n is large enough,

$$\bar{X} \doteq N(\mu, \sigma^2/n). \qquad (2.7.10)$$

(We use $\doteq$ to denote 'is approximately distributed as'.) Thus, if approximation (2.7.10) holds, probabilities associated with the sample mean $\bar{X}$ can be evaluated approximately by using normal tables. How large n has to be in order to be 'large enough' depends on how good an approximation is required and how near the underlying distribution is to a normal distribution. If the pdf of X_1 is nearly symmetric with a central peak, then values of n in excess of about 5 may well be adequate. On the other hand, where the distribution is skewed, as in the case of the exponential distribution, or where there are outliers, n may need to be in excess of 30 or more for the approximation to be useful.

You can obtain an illustration of a consequence of the central limit theorem by running program HIST. Select the default dataset and the option 'Transform'. Set the transformation to RND(1). The original data are then replaced by 47 randomly selected numbers from the uniform distribution between 0 and 1 (i.e. $U(0,1)$). (RND(1) is an instruction for generating random numbers.) On drawing a histogram using the

'Histogram' option, you will see that there is no particular pattern in the data. Select the 'Transform' option again and set the transformation to $(RND(1) + RND(1))/2$. Another 47 numbers are generated; this time they are the sample means of pairs of numbers from the $U(0,1)$ distribution. On examining the histogram, you will now see that there is a peak or peaks towards the centre. Try the 'Transform' option once more with the transformation $(RND(1) + RND(1) + RND(1) + RND(1) + RND(1))/5$, i.e. simulate sample means each based on 5 values from the $U(0,1)$ distribution. Subject to a fair degree of random variation, the shape of the histogram is now approximately that of a normal curve. You can make the comparison by choosing the 'Normal density' option. (In fact, the random variation obscures how close the pdf of $(X_1 + X_2 + X_3 + X_4 + X_5)/5$ is to a normal pdf when the X_i have a uniform distribution. In order to see this better it is necessary to simulate several hundred sample means.)

Worked example
Different subjects' reaction times to a certain type of stimulus are found to have a positively skewed distribution, with mean 0.3 sec and standard deviation 0.2 sec. One hundred randomly selected subjects are tested. Evaluate (i) the expected proportion of subjects who have a reaction time less than 0.27 sec and (ii) the probability that the sample mean reaction time of the subjects is less than 0.27 sec.

Answer. (i) The required expected proportion is equal to the probability that a randomly chosen subject has a reaction time less than 0.27 sec. We are not given sufficient information to be able to evaluate this.

(ii) Since the sample size is so large, we can almost certainly use result (2.7.10), so that the sample mean reaction time $\bar{X}$ has approximately the $N(0.3, 0.2^2/100)$ distribution. The required probability is

$$P(\bar{X} < 0.27) \simeq \Phi((0.27 - 0.3)/(0.2^2/100)^{1/2}) = \Phi(-1.5) = 0.067.$$

Another consequence of the central limit theorem is that many distributions can be approximated by normal distributions for certain ranges of their parameters. For instance, use program DISTN to show that the gamma distribution with parameters k and v is approximated by the $N(k/v, k/v^2)$ distribution for $k > 60, v > 1$. When such an approximation exists, the evaluation of probabilities is made much easier. Discrete probability distributions can also be approximated by normal distributions, but a little extra care has to be taken when applying this approximation to evaluate probabilities.

Let us investigate how binomial probabilities can be evaluated approximately using normal tables. Run program B&P and draw the $B(30,0.5)$ distribution. Select the 'Normal approximation' option. You will see that the bins are replaced by rectangles of the same heights. Since the width of each rectangle is one, the area of each is equal to the probability. You will next see the $N(15,7.5)$ pdf superimposed—it appears to be just a 'smoothed out' version of the binomial distribution. Suppose, then, that we wished to evaluate $P(X = 14)$ when $X \sim B(30,0.5)$. This is the area of the appropriate binomial rectangle, which in turn is approximately equal to the area under the normal curve between 13.5 and 14.5. In short, $P(X = 14) \simeq P(13.5 < Y < 14.5)$ where $Y \sim N(15,7.5)$. The latter probability can be evaluated by using normal tables in the usual way. You can compare the binomial probabilities with their normal approximations by selecting the 'List the probabilities' option. The resulting display is shown in figure 2.3.

This is an example of the following general result: if $X \sim B(n, p)$ and

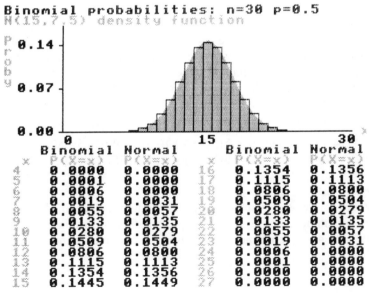

x	Binomial P(X=x)	Normal P(X=x)	x	Binomial P(X=x)	Normal P(X=x)
4	0.0000	0.0000	16	0.1354	0.1356
5	0.0001	0.0000	17	0.1115	0.1113
6	0.0006	0.0000	18	0.0806	0.0800
7	0.0019	0.0031	19	0.0509	0.0504
8	0.0055	0.0057	20	0.0280	0.0279
9	0.0133	0.0135	21	0.0133	0.0135
10	0.0280	0.0279	22	0.0055	0.0057
11	0.0509	0.0504	23	0.0019	0.0031
12	0.0806	0.0800	24	0.0006	0.0000
13	0.1115	0.1113	25	0.0001	0.0000
14	0.1354	0.1356	26	0.0000	0.0000
15	0.1445	0.1449	27	0.0000	0.0000

Figure 2.3 Comparison of the $B(30,0.5)$ and $N(15,7.5)$ distributions using program B&P. The probabilities obtained from the normal approximation are in all cases very close to the binomial probabilities.

n is large and p is not too near 0 or 1, then

$$X \doteq N(np, np(1 - p)).\tag{2.7.11}$$

Notice that the values for the mean and variance of the normal distribution are just the mean and variance of the binomial distribution. (A rule of thumb is that (2.7.11) provides reasonable approximations for most problems if $np(1 - p) > 8$.) When using (2.7.11) to evaluate approximate binomial probabilities of the form $P(c \leqslant X \leqslant d)$, remember to include the areas for the 0.5 intervals at either end of the range, i.e. express the probability as $P(c - 0.5 \leqslant X \leqslant d + 0.5)$. This is known as the *continuity correction*.

Similarly, Poisson distributions can sometimes be approximated by normal distributions. The general result is: if Y is Poisson, parameter λ, and λ is large, then

$$Y \doteq N(\lambda, \lambda).\tag{2.7.12}$$

(A rule of thumb in this case is that the approximation is likely to be reasonable if $\lambda > 8$. Check this by using program B&P.) Again, the continuity correction should be applied when evaluating probabilities.

Worked example
80% of seeds of a certain type will germinate. If 100 seeds are planted, find the probability that between 75 and 85, inclusive, will germinate.

Answer. Let X be the number of seeds germinating out of 100. Then $X \sim B(100, 0.8)$. From (2.7.11), $X \doteq N(80, 16)$. The required probability is $P(75 \leqslant X \leqslant 85)$, which we express as $P(74.5 \leqslant X \leqslant 85.5)$ for the purposes of using the normal approximation. Thus, from (2.5.7),

$$P(74.5 \leqslant X \leqslant 85.5) \simeq \Phi((85.5 - 80)/\sqrt{16}) - \Phi((74.5 - 80)/\sqrt{16})$$
$$= \Phi(1.375) - \Phi(-1.375) = 0.831.$$

Exercise
Messages arrive at a switching centre at random and at an average rate of 35 per minute. Use a normal approximation to evaluate the probability that 45 or more messages arrive in a one-minute interval.

⟩ Chapter 3

⟩ Choosing and Fitting a Probability Model

In Chapter 1 we discussed ways of displaying, summarising and examining data informally, and in Chapter 2 we saw how random variation may be described in terms of probability distributions. In this chapter we shall put these two things together by discussing how an appropriate probability distribution might be chosen for a given set of observed data. We describe this as a *model* for the data. A probability model is very useful since it allows us to make inferences about populations and so to extend our conclusions beyond the particular sample of data we have observed. We shall see examples of this in Chapters 4–6. Our first task is to decide on a family of models that is appropriate for our particular problem. The next task is to *fit* the model to the data, i.e. to choose a member of the chosen family which in some sense 'best' represents the data we have. Before moving on to carry out analyses based on this model, we should also make some attempt to *check* that our fitted model is a reasonable one. Inappropriate models may lead to inappropriate conclusions. In this chapter we shall be concerned only with informal checks. Some more formal methods will be discussed in Chapter 6.

⟩3.1 Choosing a family of models

In this section we shall consider the problem of choosing a family of models that is appropriate for the data we have observed. Consider again the thread data of Illustration 1.3, which are repeated below for ease of reference.

Illustration 3.1: thread data

The data below record the number of colouring imperfections in 50 randomly chosen lengths of thread, each 100 m long, produced by a machine.

Number of imperfections:	0	1	2	3	4	5	6	7	8
Frequency:	6	8	10	12	8	4	0	1	1.

The manufacturer of the machine claims that the mean number of faults per 100 m length is 2.

Using the methods of Chapter 1, we have already identified some features of these data through a barchart and through the sample mean and variance, which give us measures of location and scale. We saw in Chapter 1 that the sample mean of this set of data is 2.62. In order to assess whether this provides evidence against the manufacturer's claim that the population mean is 2, we must have some understanding of the random variation in the numbers of imperfections. Even if the population mean is indeed 2, we shall find some lengths with more and some lengths with fewer simply because the process is a random one. If we have a probability model for the way in which these imperfections occur then we will be better able to judge whether or not the data we have are consistent with a mean of 2.

Some specific distributions were introduced in Chapter 2, although these are in fact just a few of the many available. It will be useful to bear in mind the characteristics of the principal distributions that we met.

(1) The *binomial* distribution arises when we are counting the number of individuals who possess a certain characteristic in a random sample of individuals of fixed size.

(2) The *Poisson* distribution may be appropriate for the number of randomly occurring events observed in a fixed interval.

(3) The *normal* distribution is often a good description of data on a continuous scale. In particular, the central limit theorem leads us to expect that a normal distribution will be appropriate whenever observations are constructed by summing or averaging.

(4) The *exponential* distribution is often an appropriate model for times between events of interest, or waiting times until an event occurs.

Of course, these are by no means the only distributions which are useful in practice for modelling data. They are, however, very commonly occurring ones.

We can now consider the thread data and see if their characteristics are reminiscent of any of the distributions in the list above. The data were collected by counting the number of imperfections in 100 m lengths of thread. This fits the description of the Poisson distribution as 'the number of randomly occurring events observed in a fixed interval'. If we consider the fixed intervals to be the 100 m lengths of thread and the 'events' to be the imperfections, then this framework describes very well the way in which the thread data were collected. The Poisson distribution is therefore a prime candidate as a probability model for this set of data.

As a second example, consider again one of the exercises in Chapter 2, where a quality-control inspector in an engine assembly factory has examined 100 engines and has had to send four of these back for retuning. What is a suitable probability model for this kind of data? Here we have a group of size 100, four of which possess the characteristic 'imperfectly tuned'. The binomial distribution arises when we are interested in the number of members of a randomly selected group of individuals that possess a certain characteristic. A binomial distribution, with $n = 100$, would therefore be a natural choice of probability model for the engine tuning data.

As a further example, we might first think of the normal distribution to describe the weight of a bag of six oranges, since it arises as the sum of the weights of the individual oranges. We might first think of the exponential distribution to describe the lengths of time between accidents in a factory, since these are times between the occurrences of events. Of course, these distributions cannot describe every type of data that we will meet. They do, however, provide a very useful range of probability models.

The suggestion of an appropriate distribution is a big step forward. It is, however, only a start. For example, we have identified that the Poisson family of distributions may be appropriate for the thread data but we have not said *which* Poisson distribution. This is the topic of the next section.

⟩3.2   Fitting a model: binomial, Poisson and normal

In § 3.1 we identified the Poisson distribution as a potential probability model for the thread data, so that the probability of observing k imperfections in a 100 m length is given by

$$P(X = k) = e^{-\lambda}\lambda^k/k!.$$

λ is the mean number of imperfections that occur in a 100 m length. Further analysis of the data may require the evaluation of these Poisson probabilities, but this cannot be done until we have said which value of λ we will use. It is this process of choosing an appropriate parameter value that we refer to as *fitting* the model to the data.

Run program FREQ with the thread data and select the 'Fit a distribution' option. You will be given several distributions to choose from. Press P for Poisson. You are then invited to supply the value of the Poisson parameter λ. Since λ refers to the mean number of occurrences per 100 m length, perhaps a suitable value would be the sample mean, 2.62. In fact, this is the default value, so enter it by pressing ⟨RETURN⟩. The picture is now redrawn with the observed data represented on a relative frequency scale, in which the frequencies are each divided by the sample size, i.e. the height of each bar now records not the frequency, but the proportion of the sample that each group accounts for. Rectangles are now also superimposed on the picture to indicate the fitted Poisson probabilities. The display that should now be on the screen is illustrated in figure 3.1. Recall from § 2.3 that we should expect the relative frequencies to settle down to the true probabilities as the sample size becomes large. If we compare the relative frequencies and the probabilities here, we can see that there is reasonable agreement between them. We do not expect the probabilities to match the observed relative frequencies exactly because the latter are subject to random variation. We do, however, expect that there will be no very large discrepancies. We shall have to leave a detailed discussion of what we mean by 'large' until § 6.5. Now choose a few other values between 2 and 3 and you will see that most of these fit the data quite well. However, if a value as low as 1 is chosen, or a value near 4, then the agreement is not nearly as good.

Our hope is that the sample mean $\bar{X}$ will be near the population mean λ. Remember that the sample mean is a statistic, a random quantity derived from the data, and so it is subject to random variation. We do not expect the value of $\bar{X}$ to be exactly the true value of the unknown parameter, but we hope it will not be far away. This is an example of *estimation*. An *estimator* of a parameter is a statistic which is designed to be close to the true value of the parameter. In general, an estimator will be denoted by the parameter symbol with a 'hat' on top. In the present case we have $\hat{\lambda} = \bar{X}$, where $\hat{\lambda}$ is pronounced 'lambda hat'.

Recall the quality-control example, introduced in § 3.1, where the tuning of 100 engines is inspected. Here we have a parameter p, the probability that an engine is imperfectly tuned, and again we need to specify a value for this before further analysis can proceed. The data consist of a single observation, X say, from a binomial distribution with sample size n and probability p. n is a known value, 100 in this case, but p is an unknown parameter. Out of n engines examined, X were found to be imperfectly tuned, so perhaps a suitable estimator of p would be X/n. Since p is the population proportion of imperfectly tuned engines, the use of the sample proportion X/n as an estimator is appealing.

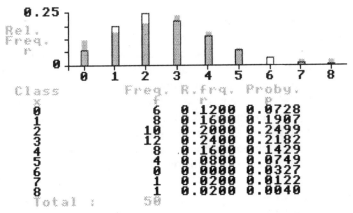

| Class | Freq. | R.frq. | Proby. |
x	f	r	p
0	6	0.1200	0.0728
1	8	0.1600	0.1907
2	10	0.2000	0.2499
3	12	0.2400	0.2182
4	8	0.1600	0.1429
5	4	0.0800	0.0749
6	0	0.0000	0.0327
7	1	0.0200	0.0122
8	1	0.0200	0.0040
Total :	50		

Figure 3.1   Fitting a Poisson distribution to the thread data in program FREQ.

Now let us consider how we might fit a normal distribution to a sample of continuous data. This is a slightly more complicated example because the normal distribution has two unknown parameters, the mean μ and the variance σ^2. However, we have sample versions of both these quantities in $\bar{X}$ and S^2, so these are natural estimators of the population mean and variance. This can be illustrated by running program HIST with the default data and selecting the 'Histogram' option, followed by the 'Normal density' option. The plotted normal density has its parameters set to the sample mean and variance. Actually, the vertical scale of this density on the picture has been altered so that it has the same area as the histogram. This makes comparisons easier to carry out. Again, having fitted the distribution, we are in a position to assess how well it describes the observed data, in this case by comparing the histogram with the scaled density function. We have already observed that there is some skewness present in the concentration data, and this is also apparent in the discrepancies between the two shapes. Perhaps a normal distribution is not a very suitable choice for this set of data. We shall return to this example in the next section.

We have now discussed how three different distributions might be fitted to observed data, in each case producing simple but intuitively appealing estimators of the unknown parameters. Now it is time to consider whether we can justify this a little more strongly. In each case we made use of a sample mean (or proportion) as an estimator of a population mean (or proportion). We know that the sample mean is a random quantity but we hope that its distribution is concentrated near the true value. At this point we can take courage from result (2.7.5), which tells us that the expected value of the sample mean is the population mean. A corresponding result applies to sample proportions. In other words, in each case the distribution of our estimator is centred on the true value. It is clearly desirable that an estimator should be centred on the quantity that we are trying to estimate. When this happens the estimator is said to be *unbiased*. Result (2.7.7) tells us that the expected value of the sample variance is the population variance, so we also know that S^2 is an unbiased estimator of σ^2, the variance parameter of the normal distribution.

In fact, from Chapter 2 we know even more than this, because result (2.7.6) tells us that $\text{var}(\bar{X}) = \sigma^2/n$, where σ^2 is the population variance. This describes the scale of the spread of the distribution of $\bar{X}$ about its mean value μ. Notice that as the sample size n increases, the variance shrinks towards zero. This is an expression of the fact that as more information becomes available, the distribution of $\bar{X}$ becomes more concentrated about μ and the accuracy of $\bar{X}$ as an estimator of μ becomes greater and greater. Roughly speaking, this means that if we add more and more observations to our sample then we will reach the correct answer in the end.

This method of estimating by matching the population mean (and, if necessary, the variance) with the sample mean (and variance) is called the *method of moments*. The name arises from the fact that the mean and variance are also called the first and second central moments of a distribution.

Exercise: checking the fit of a normal distribution
In the discussion above, the fit of a normal distribution to the concentration data was assessed by comparing the histogram with a suitably scaled

density function. This is by no means the only way of carrying out such an assessment. An alternative is to look at percentiles. Use the 'Cumulative relative frequency plot' option of program HIST to find a range of percentiles for the concentration data, say corresponding to the proportions $0.1, 0.2, ..., 0.9$. Now use program DISTN to calculate the corresponding percentage points of the normal distribution whose mean and variance are set to the sample mean and variance of the concentration data. Draw this fitted normal distribution in program DISTN and then use the 'Find $100p\%$ point' option to find the points $x_1, ..., x_9$ corresponding to the proportions $0.1, 0.2, ..., 0.9$. The fit of the normal distribution to the data can be assessed by comparing these two sets of numbers. A graphical comparison could be achieved by using program SCATTER. First use the EDITOR to create a file containing the two lists of figures and then read this file into program SCATTER. (It does not matter which variable you define to be the 'response'.) What sort of scatterplot would you expect to see if the normal distribution is a good fit?

Exercise: fitting an exponential distribution

Suppose that a sample of times between factory accidents has been collected and that the sample mean is 34.2 working days. How could we fit an exponential distribution to this set of data? The exponential probability density function is $f(x) = ve^{-vx}$, where the parameter v is greater than 0. What estimator does the method of moments suggest for v? You will need to refer to § 2.5 to find the mean of the exponential distribution.

Discussion: criticising assumptions

Consider again the quality-control example, where 100 engines from an assembly line are examined and four are found to need retuning. Earlier, we identified that a binomial distribution, with $n = 100$, would be a natural choice of probability model in this situation. We should, however, always be prepared to criticise assumptions. Are there any reasons for suspecting that the assumptions of a binomial distribution might not be satisfied?

One important question to ask is whether the engines were randomly selected for testing or whether 100 consecutive engines were used. In the latter case the observations may not be truly independent. For example, imperfect tuning may be due to slippage in the adjustment of a machine or tiredness of an operator as the end of the working day approaches.

We may then find that consecutive engines tend to have similar characteristics and so the observations are not independent. If effects of this kind are not present, we may proceed with the binomial model. However, we could in any event adopt the binomial distribution as a 'working assumption' and proceed to test any suspected defects in this model.

A summary of the formulae for estimators appropriate to the common probability models discussed in this section is provided in § 3.4.

⟩3.3   Transformations of the data

If we find reason to doubt the initial assumptions about the distributional shape of our data, we are not necessarily forced to choose another distributional family as our probability model. An alternative is to consider a *transformation* of our scale of measurement. For example, we have commented on several occasions that the concentration data are positively skewed and we may therefore be unhappy about fitting a normal distribution to them. One possibility is to consider fitting a normal distribution to some suitable transformation of the concentration values.

We can explore the effects of transformations by generating some normally distributed data in program HIST. To do this, run the program with the default set of data and select the 'Transform' option. Now enter the rather complicated expression

$$SQR(-2*LN(RND(1)))*COS(2*PI*RND(1)).$$

RND(1) is a function which simulates a value from the uniform distribution between 0 and 1. It can be shown that the expression you have just typed in generates an observation from a standard normal distribution. (For details of this and other related techniques, see *Elements of Simulation* by B J T Morgan, referred to in the Bibliography.) Since there were 47 observations in the concentration data, the computer has now generated 47 observations from a standard normal distribution. Select the 'Histogram' option, and a reasonably symmetric shape should be apparent. Of course, because of random variation, the histogram will not look exactly like a normal density but it should be unimodal and reasonably symmetric. Select the 'Transform' option and enter $X + 3$ in order to make all the observations positive. The value 3 should suffice for this but, if it does not, use a slightly higher value. We still have a sample of observations from a normal distribution, although the mean of the distribution is now larger than 0.

Now transform the data once more, this time using the transformation EXP(X), which the computer understands to represent the exponential function e^X. We no longer have normally distributed data and the histogram displayed should be strongly skewed to the right, something like figure 3.2. However, we know that if we apply the natural logarithm transformation (the inverse function of the exponential) to this sample of observations, then the resulting data will have a normal distribution because this is where we started. So, if we observe data which are strongly skewed to the right, we may be able to produce a sample whose distribution is close to normal by applying a log transformation. (You may find that when the EXP(X) transformation is applied, there are one or two outliers to the right of your picture which make the shape of the histogram rather difficult to see. A clearer picture of the majority of the data can be obtained by sorting the data and then editing out these outliers.)

Rerun program HIST with the concentration data and apply a natural logarithm transformation, which is represented on the computer by

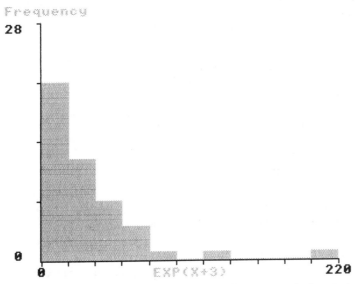

Figure 3.2 A histogram of data simulated from the standard normal distribution and transformed by the function e^{X+3}, using program HIST.

LN(X). The resulting histogram is more symmetric and, if a normal density is superimposed, plausibly normally distributed. So, a more satisfactory analysis of these data would be based on log(concentration) rather than on the concentration itself.

This simple example demonstrates the effectiveness of a suitably chosen transformation.

Illustration: square root transformation
Repeat the process of generating a normal random sample through the transformation SQR($-2*$LN(RND(1)))$*$COS($2*$PI$*$RND(1)) and again ensure that positive values are obtained by adding 3, or some other suitable small number, to each observation. Now apply the transformation $X*X$. The resulting picture should look somewhat similar to the previous one, which was skewed to the right, although you should find that the skewness is not quite so strong. So if we wish to remove skewness of this type from a sample of data, and we find that the log transformation is too strong, then a square root transformation may be a helpful alternative.

Exercise: skewness to the left
Generate another normal random sample, with mean 3, and ensure that positive values are obtained by adding another small constant if necessary. How might we transform the data to create a sample which has skewness to the left rather than to the right? Try the transformation $-1/X$. (The minus sign is present to ensure that the order of the data is unchanged. If it is omitted, the smallest original observation becomes the largest new observation, and so on.) If we now wish to remove skewness to the left from a sample, can you suggest a suitable transformation? Try it and see.

⟩3.4   Maximum likelihood estimation

The 'method of moments' described in § 3.2, where population parameters are estimated by their sample equivalents, suffers from a number of limitations. For example, the mean and variance of the Poisson distribution are both λ, so that the sample mean and variance are both unbiased method-of-moments estimators of λ. Since the sample mean and variance will not, in general, be equal, we see that one deficiency of the method is that it does not lead to unique estimators.

Another deficiency is that the method does not necessarily lead to *efficient* estimators. The efficiency of an estimator is a measure of how tightly its distribution is concentrated around the value of the parameter. Among unbiased estimators, the most efficient estimator is the one whose variance is the smallest. For example, with Poisson data it can be shown that the variance of the sample variance exceeds the variance of the sample mean. The sample mean is therefore more efficient, and so is the preferred estimator.

An important question, then, is whether there is a method of parameter estimation that will produce the 'best' estimator in every situation. Unfortunately, the answer to this question is 'no'. However, there is a method of estimation, called 'maximum likelihood', which has been found to produce good estimators in most situations and which has some very attractive properties attached to it, at least for large sample sizes.

We can use program FREQ again to provide a brief indication of what the method of maximum likelihood estimation involves. In order to keep things as simple as possible, we will consider the situation of a single observation from a Poisson distribution. In the context of the thread example, we would have only a single 100 m length. This is not a very practical situation, but it keeps the theory at a very simple level for the purposes of illustration. Select the 'New data' option, choose integer categories from 0 to 7 and enter the single observation 4 as data (i.e. in response to the prompts, enter the frequencies 0, 0, 0, 0, 1, 0, 0, 0.) As before, we can fit Poisson distributions with a variety of parameter values. This time we will consider how we might quantify the suitability of each parameter value that we try. We would expect from the previous section that a good estimate will be somewhere in the region of the sample mean, which in the present case is 4. Begin by fitting a Poisson distribution with parameter 1. Clearly, this value is not suitable because most of the probability in the fitted distribution lies near 1, whereas the only observation lies at 4. One measure of the match between the distribution displayed and the observation is the probability that the specified distribution assigns to the value 4. This probability, called the *likelihood* in this context, can be read from the columns of figures underneath the barchart. In the present case the likelihood is 0.0153.

Now fit a Poisson distribution with parameter 7. The fact that this is too large a value is again clear from the barchart because most of the fitted distribution lies above the value 4. Again this may be quantified by the probability of the outcome 4 under the model, namely 0.0912, which is again a very small value. Repeat this procedure for the parameter values 2, 3, 4, 5 and 6. You may now like to make a quick paper and pencil sketch of these likelihoods as a function of λ. (Alternatively, create a file containing the parameter values and the corresponding likelihoods. Read this file into program SCATTER, identifying the likelihood as the 'response', and a plot will be produced.) Of course, any positive real number is valid as a parameter of a Poisson distribution, so you can join up the points on your sketch to produce a smooth graph. You will find that the graph reaches a maximum near the value 4. This is reassuring since we knew from the start that 4 would be a sensible choice. What we have done is to assess each parameter value by its *likelihood*. For discrete distributions, the likelihood of a parameter value is simply the probability that the model assigns to the observed data when this parameter value is used. The *maximum likelihood estimate* is then the value of the parameter that maximises the likelihood. In this sense, the maximum likelihood estimate is the most plausible value.

This idea is not hard to generalise to the case of a random sample. Suppose X is replaced by a collection of independent observations $(X_1, ..., X_n)$. Let us consider how the argument carries through. Select the 'New data' option again and enter the three observations 2, 4 and 5. (Again, for a clear display it is best to set the smallest and largest integers to 0 and 7 and to set the frequencies to 0, 0, 1, 0, 1, 1, 0, 0.) Now fit the Poisson distribution with parameter 2. How should we quantify the fit of this distribution? From the figures beneath the barchart we can read off the probabilities of the outcomes 2, 4 and 5 as 0.2707, 0.0902 and 0.0361 respectively under the Poisson distribution with mean 2. Since we are assuming that we have a random sample, the probability of the outcome $(2, 4, 5)$ for the three observations is simply the product of the individual probabilities, i.e. $0.2707 \times 0.0902 \times 0.0361 = 0.000\,881\,5$. The multiplication rule applies because the observations are independent.

So, when we have a random sample, the likelihood function is defined by

$$L(\lambda) = P(X_1 = x_1, ..., X_n = x_n; \lambda) = P(X_1 = x_1; \lambda)...P(X_n = x_n; \lambda)$$

$$= \prod_i P(X_i = x_i; \lambda).$$

where the notation $L(\lambda)$ emphasises that we are viewing this probability as a function of λ. The notation $P(X_i = x_i; \lambda)$ for each individual probability also emphasises that λ is the current value of the parameter being used in the model. (The symbol $\prod_i$ indicates that the terms

$P(X_i = x_i; \lambda)$ are to be multiplied together.) Try a selection of parameter values, in each case noting the probabilities assigned to the three observations and multiplying these together to obtain the value of the likelihood. Again, sketch the likelihood as a function of λ. You will find that the maximum occurs at a parameter value somewhere between 3 and 4. This is again consistent with the sample mean, which takes the value 11/3.

It can be shown that for the Poisson model the maximum likelihood estimator of λ is the sample mean and, for the binomial model, the maximum likelihood estimator of p is the sample proportion X/n. In these cases, then, the maximum likelihood estimators coincide with the method-of-moments ones. However, the principle of likelihood is a very general one and it can be applied in a very wide variety of situations. Maximum likelihood can also be applied to continuous distributions, where the likelihood of a parameter value θ, based on a sample of observations $X_1, \ldots, X_n$, is defined to be the joint density function evaluated at the observations, which, by (2.7.2), is

$$L(\theta) = f(X_1; \theta)f(X_2; \theta) \ldots f(X_n; \theta).$$

$\rangle$3.5 Summary

In this chapter we discussed ideas relating to *fitting* a probability model to a set of data. The concept of an *estimator* of an unknown parameter was introduced. It was shown that the sample mean (and variance) provide good estimators of the parameters of some standard distributions. Specifically,

Distribution	Parameter(s)	Estimator(s)
Binomial	p	X/n
Poisson	λ	$\bar{X}$
Normal	μ, σ^2	$\bar{X}, S^2$
Exponential	v	$1/\bar{X}$

The estimator shown for the binomial parameter p is a function of X, a single binomial observation. If a sample $X_1, \ldots, X_m$ of binomial observations is available, then the corresponding estimator is $\Sigma_i X_i/nm$.

Plots and barcharts, with the fitted distribution superimposed, provide a simple way of *checking* whether the chosen model describes the data adequately. *Transformations* can be a very effective way of removing undesirable features such as skewness.

The principle of *maximum likelihood* provides a powerful general-purpose method of parameter estimation.

⟩ Chapter 4

⟩ Confidence Intervals

We have seen that it is often possible to describe randomly varying data by a probability distribution of a known type. The parameter, or parameters, of this distribution are frequently useful summary measures associated with the population (e.g. the population mean). Since estimators of parameters are themselves random variables, it is often unsatisfactory to represent knowledge about parameters simply by single estimates. What is also required is a measure of the precision of each estimate, i.e. a range of values within which we are fairly certain that the parameter lies. Such a range is called a *confidence interval* or *interval estimate*. We shall discuss the construction of confidence intervals for the population mean and the proportion of a population possessing a characteristic.

⟩4.1 Normally distributed data, variance known

One of the simplest cases to consider is where the data consist of a random sample of measurements $X_1, X_2, ..., X_n$ which are normally distributed, with known variance σ^2, and it is desired to find a confidence interval for μ, the mean. As an example, consider the following situation.

Illustration 4.1
An educational testing organisation has devised a simple construction task, designed to test the analytical and manipulative skills of five-year-old children. In order to calibrate this task, a random sample of 10 five-year-olds is selected and their times (in seconds) to complete the task are

recorded as follows:

52.7   60.8   40.9   39.5   42.6   47.1   35.9   59.2   66.7   48.5

It is assumed that these values come from a normal distribution with variance 100 seconds^2 (i.e. standard deviation 10 seconds). It is required to find a confidence interval for μ, the mean time to complete the task for all five-year-olds.

Recall from § 2.7 that the distribution of the sample mean, $\bar{X} = \Sigma X_i/n$, is also normal, with mean μ and variance σ^2/n. Standardising in the usual way (c.f. result (2.5.4)):

$$\frac{\bar{X} - \mu}{(\sigma^2/n)^{1/2}} \sim N(0, 1). \tag{4.1.1}$$

Notice that in this result, the only quantity on the left-hand side that we do not know, at least after collecting the data, is μ. Use program DISTN to confirm that a random variable with the $N(0, 1)$ distribution has a probability of 0.95 of taking a value between -1.96 and 1.96. (Formula (2.5.9) indicates how these values are found using table 2 of Appendix 4.) So, with probability 0.95,

$$-1.96 < \frac{\bar{X} - \mu}{(\sigma^2/n)^{1/2}} < 1.96.$$

Multiplying by $(\sigma^2/n)^{1/2}$,

$$-1.96(\sigma^2/n)^{1/2} < \bar{X} - \mu < 1.96(\sigma^2/n)^{1/2}$$

and rearranging,

$$\bar{X} - 1.96(\sigma^2/n)^{1/2} < \mu < \bar{X} + 1.96(\sigma^2/n)^{1/2}. \tag{4.1.2}$$

Since the relation (4.1.2) holds with probability 0.95, it seems reasonable to call the interval

$$(\bar{X} - 1.96(\sigma^2/n)^{1/2}, \bar{X} + 1.96(\sigma^2/n)^{1/2})$$

or   (4.1.3)

$$(\bar{X} \pm 1.96(\sigma^2/n)^{1/2})$$

(i.e. all values between $\bar{X} - 1.96(\sigma^2/n)^{1/2}$ and $\bar{X} + 1.96(\sigma^2/n)^{1/2}$) a *95% confidence interval* for μ.

We shall see how such a confidence interval is computed in the context of Illustration 4.1. Run program CONF and you will see that the times to complete the task constitute the default dataset. Select the 'Evaluate a confidence interval' option. The data points are plotted on the axis at the bottom of the screen. It is clear that a single 'best' estimate of μ will be approximately 50, but it is by no means clear how precise this estimate is. For instance, is 40 a plausible value for μ on the basis of these data?

We must commence by telling the computer what we know of the distribution of the data. It already assumes that the data are normally distributed. Type Y in response to the 'Is σ^2 known?' query. A formula similar to (4.1.3) is displayed. The computer will now take us step by step through the calculations required to compute the 95% confidence interval. First, the sample mean $\bar{X}$ is evaluated. Next, enter the value of the variance, 100, and the confidence level, 95. In response, the computer indicates that the value of z in the formula on the screen is 1.96, as we have already seen in (4.1.3). On pressing the space bar once more, the confidence interval formula is evaluated and the confidence interval for μ is indicated graphically.

You can see that the interval extends from approximately 43.2 to approximately 55.6. Notice that not all the data values are contained within the confidence interval. This is not really surprising, since the interval relates to the population mean μ, not to the behaviour of individual observations.

A 95% confidence interval has a 1 in 20 chance of not including μ in a sense to be discussed below. Suppose that we want an interval that is 99% sure to include μ. Select the 'Change the confidence level/sample size' option, retain the sample size at 10 and set the confidence interval to 99. The values of $\bar{X}$ and σ^2 are unchanged, but the value of z is increased to about 2.576. This value is chosen because there is a probability of 0.99 that an $N(0, 1)$ random variable lies between -2.576 and 2.576. (You can run program DISTN to confirm this; note that 2.576 is $z_{0.995}$, the 99.5% point of the $N(0, 1)$ distribution, as indicated by formula (2.5.9).) Pressing any key, you will see that the 99% confidence interval is about (41.3, 57.5). Notice that it is wider than the 95% confidence interval. This is the penalty of being that much more sure of including μ.

Exercise

What value of z should be chosen for a 90% confidence interval? What is the 90% confidence interval? Answer these questions firstly by using tables (or program DISTN) and confirm your answers by using program CONF.

Let us now investigate what might be expected to happen if the sample size were increased. Select the 'Change the confidence level/sample size' option and set the sample size to 50 and the confidence level to 95. The computer assumes that the computed value of $\bar{X}$ and the known value of σ^2 are as before. The value of z is 1.96, as for the previous 95% confidence interval, but the width of the interval has decreased since n is larger than before. More intuitively, the smaller width reflects the fact that the more data that are available, the more precise is the estimate of μ. Indeed, one criterion used for deciding the sample size prior to conducting a survey or a series of experiments is to specify the width of a confidence interval and a value for the variance.

Exercise
Use program CONF to calculate the 95% confidence interval that would be produced for a sample of size 1000, assuming that the sample mean takes its present value. Now use trial and error to decide how large a sample is required in order for the width of the 95% confidence interval to be 2.0 (i.e. the interval is of the form $\bar{X} \pm 1.0$). (A formula for calculating this sample size is given at the end of this section.)

Let us now consider in rather more detail the sense in which the interval given by (4.1.3) can be described as a 95% confidence interval. Recall, for instance, that the 95% confidence interval for the task times data was approximately (43.2, 55.6). It is not true simply to say that the probability that μ lies between 43.2 and 55.6 is 0.95, since μ is a constant, not a random variable (albeit an unknown constant). Either μ lies between 54.2 and 55.6, or it does not, i.e. the probability of its lying in the confidence interval is either 1 or 0.

To see how this dilemma is resolved, select the 'Set up simulation' option of program CONF. Set n, the sample size, to 10 and type N to show that we are going to use the confidence interval formula for normal data. Set $\mu = 50$ and $\sigma^2 = 100$, i.e. we are simulating from a distribution similar to that for the data of Illustration 4.1. Indicate that the value of σ^2 is known and set the confidence level to 95%. Finally, type N to indicate that normally distributed data are to be simulated. A display similar to that shown in figure 4.1 will be produced.

You will see that 40 confidence intervals are constructed. What is happening is that the distribution of $\bar{X}$ (i.e. $N(50, 100)$) is simulated 40 times and each time formula (4.1.3) is used, with $\bar{X}$ set equal to the simulated value, to construct the confidence interval. You will see that most of the confidence intervals are white, while some may be shaded (or coloured). The former are those intervals that include 50, the value of the mean μ, i.e. these are the 'satisfactory' intervals. Of course, we only know that they are satisfactory because we defined the simulation and so know the value of μ. In practical applications, only one confidence interval is available and there is no way of knowing for sure whether it is satisfactory.

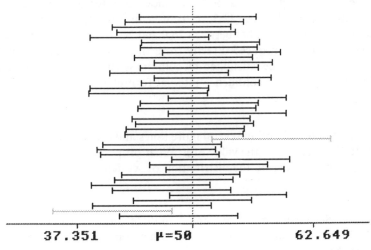

95% confidence intervals for μ
$\bar{X} \pm z\sqrt{(\sigma^2/n)}, n=10$
38/40 ints include μ

37.351 $\mu=50$ 62.649

Figure 4.1 40 confidence intervals based on data simulated from the $N(50,100)$ distribution. The confidence level is 95% and the variance is assumed to be known.

A count is kept of the number of satisfactory intervals. Select the 'Simulate' option a number of times in order to obtain further simulated intervals. Note that a cumulative count is kept of the number of simulations and of the percentage of satisfactory intervals. You will find that the proportion of satisfactory intervals tends to 0.95 as the number of simulations increases. So, a randomly chosen 95% confidence interval will include the population mean with probability 0.95. A statement that a particular interval, (43.2, 55.6), say, is a 95% confidence interval is not, therefore, a statement about the probability distribution of μ, but a statement about the method by which the confidence interval was constructed. The 'method', i.e. formula (4.1.3), has a 95% success rate. (More formally, the interval defined by (4.1.3) is called a *random interval*, and has a probability of 0.95 of including μ. Any particular interval evaluated from formula (4.1.3) is just a single observation of this random interval.)

The distinction between probabilities associated with random variables and confidence intervals may seem somewhat subtle, but it does mean that confidence intervals cannot be operated on in the same way as ordinary probability statements. For example, if in one experiment a 95% confidence interval for μ is found to be (22.3, 26.4), and in another, independent experiment a 95% confidence interval for the same parameter is found to be (22.3, 26.0), it cannot be claimed that there is 'zero confidence' that μ lies between 26.0 and 26.4. An overall confidence interval cannot be deduced directly from the two individual confidence intervals, but must be calculated by combining the two sets of data and using an appropriate confidence interval formula.

Exercise

Use the 'Simulate' option several more times and examine those intervals that fail to include 50, the value of μ. What proportion seem to be larger than μ? By considering the distribution of $\overline{X}$, determine why this is so.

Exercise

Use the 'Set up a simulation' and 'Simulate' options to check on the behaviour of confidence intervals with other confidence levels (e.g. 90% and 99%).

Mathematical details

The data consist of a random sample X_1, X_2,..., X_n from the $N(\mu, \sigma^2)$ distribution, where σ^2 is known and it is desired to find a $100c\%$ confidence interval for μ, where c is a specified value between 0 and 1, exclusive. Then the confidence interval is

$$(\overline{X} - z_{(1+c)/2}(\sigma^2/n)^{1/2}, \overline{X} + z_{(1+c)/2}(\sigma^2/n)^{1/2})$$

or
$$(\overline{X} \pm z_{(1+c)/2}(\sigma^2/n)^{1/2})$$

(4.1.4)

where $\overline{X}$ is the sample mean and $z_{(1+c)/2}$ is the $(1 + c)/2$ point of the $N(0, 1)$ distribution, given in table 2 of Appendix 4.

We shall now confirm formula (4.1.4). From formula (2.5.9), the probability that an $N(0, 1)$ random variable lies between $-z_{(1+c)/2}$ and $z_{(1+c)/2}$ is equal to c, the confidence level. Therefore, using result (4.1.1),

$$-z_{(1+c)/2} < \frac{\bar{X} - \mu}{(\sigma^2/n)^{1/2}} < z_{(1+c)/2}$$

with probability c. Rearranging this,

$$\bar{X} - z_{(1+c)/2}(\sigma^2/n)^{1/2} < \mu < \bar{X} + z_{(1+c)/2}(\sigma^2/n)^{1/2}$$

with probability c, confirming (4.1.4).

Exercise
In a small survey of students at a certain college living in private rented accommodation, the following amounts of weekly rents were reported:

£24, £33, £27, £36, £25, £30, £32, £28, £21, £31, £31, £29, £20.

Find a 90% confidence interval for the mean weekly rent per student if the standard deviation is known to be £4. (Answer first by using formula (4.1.4) and table 2. Check your answer by using program CONF, and program DISTN if necessary. (New data may be entered in program CONF by using the 'Change the data' option.) Note that two implicit assumptions are being made: (i) random sampling was adopted in the survey; (ii) the data follow a normal distribution. We shall see in § 4.3 that the latter assumption is less critical.)

It was mentioned earlier that the width of a confidence interval might be a determinant of sample size. For example, suppose that from a preliminary survey, it is known that (i) the data are normally distributed, and (ii) their variance is approximately σ_0^2. Furthermore, it is required that the $100c\%$ confidence interval for μ should be of the form $\bar{X} \pm d$, where c and d are specified values, indicating the required precision of the full survey. Then, from (4.1.4), the sample size of the full survey must satisfy

$$d = z_{(1+c)/2}(\sigma_0^2/n)^{1/2}.$$

Solving for n, we obtain

$$n = (z_{(1+c)/2})^2\sigma_0^2/d^2. \tag{4.1.5}$$

Since σ_0^2 is an estimated value, formula (4.1.5) should be thought of as providing a guide to the order of magnitude of n, rather than indicating its exact value.

Exercise (*continued*)

Suppose that in the student rent survey, it was required to estimate the mean rent per student to within £1 with 95% confidence. How many more students need to be included in the survey? (Use formula (4.1.5) to find the required sample size. Confirm your answer by using program CONF.)

⟩4.2 Normally distributed data, variance unknown

We are going to relax the assumption that the variance σ^2 is known, and see how confidence intervals are now computed. Certainly it is important that we drop this assumption, since it is rare in practice for it to hold. We again assume that we are dealing with a random sample of normally distributed data, $X_1, X_2, ..., X_n$, and that a confidence interval for μ is required.

The confidence interval formula (4.1.3) derived in the last section can no longer be used since it involves σ^2. We therefore need to go back to basics and rederive a confidence interval appropriate to this case. Recall that the starting point in § 4.1 was the result

$$\frac{\bar{X} - \mu}{(\sigma^2/n)^{1/2}} \sim N(0, 1).$$

Since σ^2 is not known, it seems sensible to estimate it from the sample variance

$$S^2 = \frac{1}{n-1} \left(\sum (X_i - \bar{X})^2 \right) = \frac{1}{n-1} \left(\sum X_i^2 - n\bar{X}^2 \right) \qquad (4.2.1)$$

introduced in § 1.2. (Note that we need to assume that $n > 1$ in order to obtain an estimate of σ^2.) We therefore need to know the distribution of

$$\frac{\bar{X} - \mu}{(S^2/n)^{1/2}}. \qquad (4.2.2)$$

At first sight this looks rather a daunting task, since the distribution might depend on μ, σ^2 and n. Fortunately, it can be shown that the distribution in fact only depends on n. (The derivation of this result may be found in more advanced texts on probability theory.) The distribution is called the 't distribution with $n - 1$ degrees of freedom', or t_{n-1} for short. It is also sometimes referred to as the Student's t distribution, 'Student' being the *nom de plume* of its discoverer, W S Gossett.

In order to get a feel for what the various t distributions look like, run program DISTN. First draw the $N(0, 1)$ distribution and then select the 'Student's t' option. Set v, the number of degrees of freedom, equal to 1. You will see that the t_1 distribution is symmetric about 0, but is rather more spread out than the $N(0, 1)$ distribution. Draw the t_2, t_5 and t_{30} distributions also. Notice that as v increases, so the t_v distribution approaches the $N(0, 1)$ distribution. By the time that $v = 30$, the two distributions are very close, apart from the extreme tails. The quantity v, the number of degrees of freedom, is just the parameter of the distribution. It must take positive, integer values.

In order to evaluate the confidence intervals, it is necessary to find $100p\%$ points of t distributions. We shall denote the $100p\%$ point of the t_v distribution by $t_{v,p}$, i.e. if the random variable Y has the t_v distribution, then $P(Y \leqslant t_{v,p}) = p$. In table 3 of Appendix 4, $t_{v,p}$ is tabulated for various values of p and v. Since all the t distributions are symmetric about 0, $t_{v,1-p} = -t_{v,p}$ and, in particular, $t_{v,0.5} = 0$ for all v.

Exercise
Use table 3 to evaluate $t_{4,0.95}$, $t_{30,0.99}$ and $t_{57,0.995}$. (You will need to interpolate in the last case.) Confirm your answers by using program DISTN.

Having been introduced to the t distribution, let us now return to the question of constructing a confidence interval for μ when σ^2 is unknown. We were considering the expression (4.2.2) and had noted the result that for $n > 1$,

$$\frac{\bar{X} - \mu}{(S^2/n)^{1/2}} \sim t_{n-1}. \tag{4.2.3}$$

Suppose that we want to find a 95% confidence interval for μ. From result (4.2.3), with probability 0.95,

$$t_{n-1,0.025} < \frac{\bar{X} - \mu}{(S^2/n)^{1/2}} < t_{n-1,0.975}$$

and, since $t_{n-1,0.025} = -t_{n-1,0.975}$, it follows that

$$-t_{n-1,0.975} (S^2/n)^{1/2} < \bar{X} - \mu < t_{n-1,0.975} (S^2/n)^{1/2}$$

with probability 0.95. Rearranging this inequality, we obtain

$$\bar{X} - t_{n-1,0.975} (S^2/n)^{1/2} < \mu < \bar{X} + t_{n-1,0.975} (S^2/n)^{1/2}$$

with probability 0.95. We have established that a 95% confidence interval for μ is

$$(\bar{X} - t_{n-1,0.975} \, (S^2/n)^{1/2}, \bar{X} + t_{n-1,0.975} \, (S^2/n)^{1/2})$$

or   (4.2.4)

$$(\bar{X} \pm t_{n-1,0.975} \, (S^2/n)^{1/2}).$$

We shall demonstrate the use of this formula in the context of Illustration 4.1, with the modification that the value of σ^2 is no longer assumed known. Run program CONF, using the data of Illustration 4.1, the default dataset. Select the 'Evaluate a confidence interval' option and indicate that σ^2 is not known. The formula for the confidence interval is displayed near the top of the screen and the sample mean is evaluated. On pressing the space bar, the value of S^2, evaluated from formula (4.2.1), is shown. Press the space bar again and enter '95' as the confidence level. The t value, the 97.5% point of the t_9 distribution, is indicated; use table 3 to check this value. All the elements of the confidence interval formula (4.2.4) have now been found, so it remains to put them together by pressing the space bar.

Use the 'Change sample size/confidence level' option to reset the sample size to 50, whilst retaining the 95% confidence level. You will see that a new value is found for t, since the number of degrees of freedom is now 49, rather than 9 as before. Note also that this new t value is close to 1.96, the corresponding value for the $N(0, 1)$ distribution. The confidence interval is next evaluated with $\bar{X}$ and S^2 taking their previous values. You will see that the length of the confidence interval is less than half the length of the interval based on 10 observations.

Exercise
Approximately what sample size would be required for the length of the confidence interval to be 4?

Exercise
Returning to the $n = 10$ case, what would the t value be for a 99% confidence interval? What is the 99% confidence interval? (Work out the answers to these questions for yourself before confirming them by using program CONF.)

The probabilistic interpretation of the 95% confidence interval formula (4.2.4) is as for the variance known case, i.e. it is a random interval with probability 0.95 of including μ, whatever the value of μ. Once particular values have been assigned to $\bar{X}$ and S^2, so that a numerical confidence interval is obtained, we have a particular outcome of a '95% accurate process'. In order to demonstrate this, select the 'Set up simulation' option, set $n = 5$, type N, set $\mu = 50$, $\sigma^2 = 100$ and type N to show that the value of σ^2 is not assumed known for the purposes of evaluating confidence intervals. Set the confidence level to 95% and type N. Forty simulations of sets of five observations from the $N(50, 100)$ distribution are made and, for each simulation, the confidence interval given by (4.2.4) is evaluated.

The display of the confidence intervals is similar to that for the 'σ^2 known' case discussed in § 4.1. The confidence intervals vary; most include 50, the mean, whilst a small number, darker shaded or coloured, may miss it. Select the 'Simulate' option a number of times to demonstrate that the proportion of intervals including 50 settles down to 0.95.

One thing that is different from the 'σ^2 known' case is that the lengths of the intervals are variable. Indeed, the longest intervals may well be three or more times longer than the shortest intervals. This is because the width of the intervals partly depends on the value of S^2, the sample variance, and this quantity is a random variable. Frequently, confidence intervals fail to include μ not because the value of $\bar{X}$ is all that far from μ but because the value of S^2 is low and the interval quite short. We shall return to this point in the next section.

Mathematical details
The data consist of a random sample $X_1, X_2, ..., X_n$ from the $N(\mu, \sigma^2)$ distribution, where μ and σ^2 are unknown, and it is desired to find a $100c\%$ confidence interval for μ, where c is a specified value. Such an interval is, then,

$$(\bar{X} - t_{n-1,(1+c)/2} (S^2/n)^{1/2}, \bar{X} + t_{n-1,(1+c)/2} (S^2/n)^{1/2})$$

or   (4.2.5)

$$(\bar{X} \pm t_{n-1,(1+c)/2} (S^2/n)^{1/2})$$

where $\bar{X}$ is the sample mean, S^2 is the sample variance and $t_{n-1,(1+c)/2}$ is the $(1 + c)/2$ point of the t_{n-1} distribution. (The quantity $(S^2/n)^{1/2}$, which estimates the standard deviation of $\bar{X}$, is known as the *standard error* of the mean.)

The derivation of this result is along similar lines to the derivation of the 'z interval' given at the end of § 4.1, with the modification that result (4.2.3) is used instead of (4.1.1). Note that since the t_v and $N(0, 1)$

distributions are very close for $v > 80$, it is sometimes acceptable to use formula (4.1.4), with σ^2 replaced by S^2, instead of (4.2.5) when the sample size exceeds 80.

As we saw in § 4.1, the confidence interval formula (4.1.4) can be used to obtain an indication of the appropriate sample size. Interval (4.2.5) is not so easy to use for this purpose, since the value of $t_{n-1,(1+c)/2}$ depends on the sample size. For most purposes it is adequate to use formula (4.1.5), even where the value of σ^2 is guessed rather than known.

Exercise

An industrial pharmacist wishes to determine the mean time for absorption of a certain drug by the body. Eight randomly chosen subjects are given the drug and the following times (in minutes) for the concentration of the drug in the blood to reach a certain level are found:

$$36, \quad 29, \quad 34, \quad 33, \quad 38, \quad 28, \quad 30, \quad 33.$$

Find a 99% confidence interval for the mean absorption time. (Use formula (4.2.5) and confirm your answer by using program CONF.)

⟩4.3   Normal approximations for non-normal data

One apparent limitation of the development of confidence intervals so far is the assumption that the data come from a normal distribution. In fact, it turns out that the confidence interval formulae (4.1.4) and (4.2.5) very often provide perfectly adequate approximations even when the data come from a non-normal distribution. We shall examine this claim in the first part of this section and determine conditions under which the approximation is satisfactory.

It also turns out that the normal approximation to the binomial and Poisson distributions allows approximate confidence intervals to be derived for the parameters of these distributions. We shall discuss the binomial case in the second part of this section and give results concerning the Poisson case at the end.

Suppose that we have a random sample $X_1, X_2, ..., X_n$ from a distribution with mean μ and variance σ^2, where neither the form of the distribution nor the values of μ and σ^2 are known, and we require to find a confidence interval for μ. The confidence interval of § 4.2 was deduced from the distribution of $\bar{X}$, which was normal. But the central limit theorem (see page 57) asserts that the distribution of $\bar{X}$ is approximately normal, whatever the underlying distribution, as long as n is sufficiently large. It therefore seems reasonable to hope that (4.2.5) will provide an approximate $100c\%$ confidence interval for μ for non-normal data, for large n. The practical question is how large n has to be for the approximation to be good. We shall try to indicate a general answer to this question by simulating from two particular non-normal distributions.

The first distribution that we shall try is the uniform distribution between 32.68 and 67.32, i.e. the $U(32.68, 67.32)$ distribution. The end points have been chosen so that $\mu = 50$ and $\sigma^2 = 100$, in order to ensure consistency with the previous simulations that we have tried. Run program DISTN and draw the $N(50, 100)$ and $U(32.68, 67.32)$ distributions. Note that the uniform distribution differs from the normal in not having a central peak or tails. Now run program CONF and select the 'Set up simulation' option. Set $n = 10$, normal confidence intervals, $\mu = 50$, $\sigma^2 = 100$, σ^2 unknown and the confidence level to 95%. You will see that the formula for the t-confidence interval is displayed. This will be used to evaluate the simulated confidence intervals. Select the uniform distribution for the simulations.

The usual type of display is produced. Each interval is based on a simulated sample of 10 observations from the $U(32.68, 67.32)$ distribution. Select the 'Simulate' option a number of times. You should find that the proportion of intervals including 50, the mean of the uniform distribution, does settle down to about 0.95, confirming that formula (4.2.5) is appropriate for this distribution, at least for this sample size.

Exercise
Use program CONF to investigate how small the sample size n has to be before formula (4.2.5) is unacceptable. Investigate whether formula (4.2.5) provides approximate 99% confidence intervals for the same range of values of n.

We can conclude from the investigation so far that if the distribution of the data differs from the normal in the direction of the uniform distribution, then formula (4.2.5) provides adequate approximate confidence intervals for μ for all but the smallest values of n.

We shall now try a distribution with the same shape as the gamma distribution with $k = 2$ and $v = 1$. Run program DISTN, draw the gamma distribution, $k = 2$, $v = 1$, and the $N(2, 2)$ distribution (i.e. the normal distribution with the same mean and variance as the gamma distribution). You will see that the gamma distribution differs from the normal in being skewed to the right; it is more peaked than the normal distribution and has a 'heavier right tail', that is to say its probability density function tends to zero more slowly. This means that most values from the gamma distribution will tend to cluster together, but the occasional value will be much larger than anything likely to be generated by the normal distribution.

Run program CONF and select the 'Set up simulation' option. Set $n = 5$, normal confidence intervals, $\mu = 50$, $\sigma^2 = 100$, σ^2 unknown and the confidence level to 95%. Select the skew distribution for the simulation. The simulated intervals are based on samples of five observations from a scaled and shifted gamma distribution. (In fact, values simulated from the gamma, $k = 2$, $v = 1$, distribution are each multiplied by $\sqrt{50}$ and added to $50 - 2\sqrt{50}$ to ensure that $\mu = 50$ and $\sigma^2 = 100$.) Repeat the simulation by selecting the 'Simulate' option a number of times.

You should find that about 91% of the intervals include μ ($= 50$), i.e. formula (4.2.5) no longer seems to be adequate. Notice why so many intervals fail to include μ. It is usually because the sample mean is smaller than 50 and the length of the interval is much shorter than average. The reason for the failure of most intervals is, therefore, the clustering of points near the mode (which is less than μ), rather than very large values 'dragging' the interval too far to the right.

Exercise
Rerun the simulation, but this time specify σ^2 as known so that the z-confidence interval (4.1.4) is used. Show that for skewed data, a knowledge of the value of σ^2 allows confidence intervals based on the normal distribution to be used for smaller sample sizes than are required when σ^2 is unknown.

Now let us see whether a larger sample size leads to a more satisfactory approximation from formula (4.2.5). Select the 'Set up simulation' option, set $n = 25$, normal confidence intervals, $\mu = 50$, $\sigma^2 = 100$, σ^2 not known, the confidence level to 95% and the skew distribution. After selecting the 'Simulate' option a number of times, you should find that the proportion of confidence intervals including μ is now about 0.95, i.e. the approximation does now seem to be reasonable, although there is still some tendency for short intervals to fall to the left of μ.

Exercise
What is the smallest value of n such that the 95% t-interval appears to be adequate? Can we get away with smaller values for 90% intervals?

Confirm that larger sample sizes are required for 99% intervals.

In conclusion, when data are skewed, larger sample sizes are required to ensure that the t-interval formula (4.2.5) is adequate. Typically, if a histogram shows skewness to the extent of the gamma, $k = 2$, $v = 1$, distribution a sample size in excess of approximately 25 would be required for a 95% confidence interval. More extreme skewness, or a larger confidence level, would require a larger minimum sample size. When the sample size is not sufficiently large for a normal confidence interval to be used directly, one procedure is to transform the data so that the resulting distribution is not so skewed. (Taking logarithms or square roots are commonly recommended transformations for data which are skewed to the right; see § 3.3.) The appropriate normal confidence interval, (4.1.4) or (4.2.5), can then be evaluated from the transformed data. However, the resulting interval relates to the mean of the distribution of the logged data.

Binomial data
Recall that the binomial distribution often arises when it is desired to estimate the proportion p of a population possessing a characteristic on the basis of a random sample. The following is an example of such a situation.

Illustration 4.2
A new road has been proposed in a certain town. Opinion about the desirability of the road is divided, so the residents' association undertakes a survey. 50 people are selected at random from the voting list and asked their opinion. It is found that 30 of them say that they are in favour of the new road. It is required to find a 95% confidence interval for p, the proportion of *all* people on the voting list who are in favour of the new road.

In this case X, the number of people in the survey reporting that they are in favour of the road, has the $B(50, p)$ distribution and the observed value of X is 30. (Recall from § 2.4 that, in general, X has the $B(n,p)$ distribution if it is the number of individuals in a random sample of n individuals possessing a characteristic, where p is the probability that each individual possesses the characteristic.) Also, if $np(1 - p)$ is 'large' then from result (2.7.11),

$$X \doteq N(np, np(1 - p)).$$

(Recall that $\doteq$ stands for 'is approximately distributed as'.) By the rules for normally distributed random variables,

$$X/n \doteq N(p,\ p(1-p)/n)$$

i.e.

$$\frac{X/n - p}{[p(1-p)/n]^{1/2}} \doteq N(0, 1). \qquad (4.3.1)$$

Compare result (4.3.1) with (4.1.1). In a similar manner to the derivation of (4.1.2) it can be shown that, with approximate probability 0.95,

$$X/n - 1.96[p(1-p)/n]^{1/2} < p < X/n + 1.96[p(1-p)/n]^{1/2}. \qquad (4.3.2)$$

(Recall that a $N(0, 1)$ random variable has a probability of 0.95 of lying between -1.96 and 1.96.) We cannot, however, use $(X/n \pm 1.96 [p(1-p)/n)]^{1/2})$ as a 95% confidence interval because p is unknown. We therefore further approximate by replacing p, the population proportion, by X/n, the sample proportion. Thus an approximate 95% confidence interval for p is

$$(X/n - 1.96[(X/n)(1 - X/n)/n]^{1/2}, X/n + 1.96[(X/n)(1 - X/n)/n]^{1/2})$$

or   (4.3.3)

$$(X/n \pm 1.96[(X/n)(1 - X/n)/n]^{1/2}).$$

We shall see how this formula is used in the context of Illustration 4.2. Run program CONF and select the 'Change the data' option. Type 'B' to indicate that binomial data are being entered and set $n = 50$ and $X = 30$. Select the 'Evaluate a confidence interval' option and you will see that the sample proportion in favour of the new road, $X/n = 0.6$, is evaluated and a formula similar to (4.3.3) is shown. Press the space bar and set the confidence level to 95%. The computer shows that the appropriate z value is 1.96. Another press of the space bar shows that the 95% confidence interval for p is approximately $(0.464, 0.736)$, i.e. the confidence interval for the percentage is 46.4–73.6%.

Exercise
This interval is very wide. In particular, it includes values less than 50%, indicating that there is a fair chance that only a minority of the voters are in favour of the new road. Use the 'Change the confidence interval/ sample size' option to see how many people need to be sampled in order

that the 95% confidence interval only includes values greater than 0.5, assuming that X/n remains at 0.6. (We shall discuss this type of problem in more detail at the end of the section.)

Exercise
How does formula (4.3.3) need to be modified for 99% confidence intervals? Evaluate an approximate 99% confidence interval for p for Illustration 4.2 by using tables and check your answer by using program CONF.

We noted that it was necessary to make a number of approximations in deriving the 95% confidence interval formula (4.3.3). A rule of thumb for deciding whether this formula is adequate is to check that

$$X(1 - X/n) > 8. \qquad (4.3.4)$$

We shall use simulation to examine when (4.3.3) is satisfactory. Select the 'Set up simulation' option of program CONF, set $n = 50$, the binomial distribution, $p = 0.6$ and confidence level 95%, i.e. we are going to simulate a situation similar to Illustration 4.2. You will see that 40 confidence intervals are drawn, corresponding to 40 values of X simulated from the $B(50, 0.6)$ distribution and the use of formula (4.3.3). Select the 'Simulate' option a number of times to check that the proportion of intervals including 0.6 is approximately 0.95.

Exercise
Check that interval (4.3.3) is an approximate 95% confidence interval for the following values of n and p: (i) $n = 40$, $p = 0.5$; (ii) $n = 100$, $p = 0.3$.

Exercise
Examine the behaviour of interval (4.3.3) when condition (4.3.4) is violated, e.g. set $n = 50$, $p = 0.1$ or $n = 10$, $p = 0.5$.

Mathematical details
   (i) The data consist of a random sample of observations $X_1, X_2, ..., X_n$ from a distribution of unknown type, with mean μ and variance σ^2 which are also unknown. Then, if n is large enough, an approximate $100c\%$

confidence interval for μ is given by

$$(\bar{X} - t_{n-1,(1+c)/2} \, (S^2/n)^{1/2}, \; \bar{X} + t_{n-1,(1+c)/2} \, (S^2/n)^{1/2})$$

or   (4.3.5)

$$(\bar{X} \pm t_{n-1,(1+c)/2}(S^2/n)^{1/2})$$

where $\bar{X}$ and S^2 are the sample mean and variance and $t_{n-1,(1+c)/2}$ is the $(1 + c)/2$ point of the t_{n-1} distribution. For data which are not skewed, the approximation is likely to be reasonable for n larger than 5, say, if the data are close to normal, or n larger than 15, say, if the data are markedly non-normal. For data which are strongly skewed, or contaminated with outliers, much larger values of n may be required. Furthermore, larger sample sizes are required for high confidence levels (99%, 99.9%, etc) than for lower confidence levels (90% or 95%).

(ii) The data consist of a single observation X from the $B(n, p)$ distribution, where n is known and p is unknown. Then if inequality (4.3.4) is satisfied, an approximate $100c\%$ confidence interval for p is

$$(X/n - z_{(1+c)/2}[(X/n)(1 - X/n)/n]^{1/2},$$
$$X/n + z_{(1+c)/2}[(X/n)(1 - X/n)/n]^{1/2})$$

or   (4.3.6)

$$(X/n \pm z_{(1+c)/2} \, [(X/n)(1 - X/n)/n]^{1/2})$$

where $z_{(1+c)/2}$ is the $(1 + c)/2$ point of the $N(0, 1)$ distribution. If several independent observations $X_1, X_2, ..., X_m$ are available and X_i comes from the $B(n_i, p)$ distribution, then (4.3.6) can be used with $X = \Sigma X_i, n = \Sigma n_i$.

(iii) The data consist of the independent observations $X_1, X_2, ..., X_n$ from the Poisson distribution with mean λ. Then if $\Sigma X_i > 8$, say, an approximate $100c\%$ confidence interval for λ is

$$(\bar{X} - z_{(1+c)/2} \, (\bar{X}/n)^{1/2}, \; \bar{X} + z_{(1+c)/2} \, (\bar{X}/n)^{1/2})$$

or   (4.3.7)

$$(\bar{X} \pm z_{(1+c)/2} \, (\bar{X}/n)^{1/2}).$$

Indications of how intervals (4.3.5) and (4.3.6) are derived have already been provided in the text. Recall from § 2.7 that, when λ is large, the Poisson distribution can be approximated by the normal distribution. In fact, if $\Sigma X_i > 8$, say,

$$\sum X_i \doteq N(n\lambda, \, n\lambda)$$

from which it follows that

$$(\bar{X} - \lambda)/(\lambda/n)^{1/2} \div N(0, 1).$$

Interval (4.3.7) can now be derived in much the same manner as the approximate binomial confidence interval. Note that $(\lambda/n)^{1/2}$ is approximated by $(\bar{X}/n)^{1/2}$.

Exercise
An experiment is conducted to see whether a die is loaded in favour of the score 6. In 200 throws, 6 is obtained 40 times. Find a 90% confidence interval for p, the probability of a 6, by using (4.3.6) and confirm your answer by using program CONF.

Exercise
Calls to an emergency lift breakdown service occur at random, at an average rate of λ per night. In five nights, a total of 32 calls are made. Find a 95% confidence interval for λ.

We have already seen that one method of determining the sample size in advance of performing an experiment or survey is to specify the width of a confidence interval. In the case of data with an unknown distribution, but where the distribution of $\bar{X}$ can be assumed to be approximately normal, the method described at the end of § 4.1 is appropriate. For binomial data, if it is required that a 95% confidence interval for p should be of the form $(X/n) \pm d$, where d is specified, then inequality (4.3.2) suggests that

$$d = 1.96[p(1 - p)/n]^{1/2}$$

i.e.

$$n = 1.96^2 p(1 - p)/d^2. \qquad (4.3.8)$$

An estimate of the required value of n can then be obtained from (4.3.8) after replacing p by a prior estimate of its value. (Set $p = 0.5$ if it is not possible to make a prior estimate. This gives the factor $p(1 - p)$ its highest possible value.)

The result corresponding to (4.3.8) for Poisson data is

$$n = 1.96^2 \lambda/d^2. \qquad (4.3.9)$$

In order to use (4.3.9) to determine an appropriate sample size, λ must be replaced by a prior estimate. For both binomial and Poisson data,

'1.96' must be replaced by $z_{(1+c)/2}$ if the required precision is expressed in terms of a $100c\%$ confidence interval, where $c \neq 0.95$.

Exercise
In the lift breakdown exercise, about how many nights' data would be required if a 95% confidence interval for λ was to have a width of approximately 0.5 (i.e. $d = 0.25$)?

〉 Chapter 5

〉 Statistical Tests: Principles

Experiments and surveys are often carried out with the objective of testing a theory, or *hypothesis*, about the nature of the process under investigation. If some acceptable probability model can be found, then the hypothesis of interest may be characterised by the value(s) taken by one or more parameters of the model. When the data are subject to random variation, it will usually be the case that there is no procedure that will invariably identify whether or not a hypothesis is correct. What is then required is a procedure, or *hypothesis test*, for weighing up the evidence for or against the hypothesised value(s) of the parameter(s). This procedure will take into account the chances of making an incorrect decision.

In this chapter we shall explore some of the basic ideas of hypothesis testing in the relatively simple contexts of a single sample of data from a normal distribution with known variance, and from a binomial distribution. Further tests, with wider applicability, will be discussed in the next chapter.

〉5.1 Hypothesis Tests

Consider the following situation.

Illustration 5.1
A manufacturer of tyres is considering a change to the method of manufacture of his product. He knows that the distribution of life length of his present tyres, measured in 1000s of miles until they are worn out, is normal, with mean 40 and variance 9. 10 tyres made using the new

method are to be tested and their life length ascertained. On the basis of the results, he wishes to decide whether the mean life length has changed. (He assumes that the life length distribution will continue to be normal, with variance 9.)

One hypothesis of interest in this case is that the proposed change is ineffective, i.e. μ, the mean life length after making the proposed change, is still 40. Another hypothesis is that the change has led to an improved life length, i.e. $\mu > 40$. On the basis of the evidence of the life lengths of the experimental tyres, he must decide whether to go ahead with the change. Notice that both hypotheses refer to the population mean, while the data he obtains from the experiments will only constitute a sample of values. The results of the experiment, being subject to random variation, will not, therefore, point unequivocally to one hypothesis or the other. The best that can be done when deciding on which hypothesis to go along with is to use a method of choice which limits the chances of making a mistake in some way.

Before pursuing this particular example further, let us define the general type of situation under investigation in mathematical terms. We suppose that our random sample consists of values $X_1, X_2, ..., X_n$, which are known to come from a normal distribution. Furthermore, the variance σ^2 is known but the mean μ is not. There are two hypotheses relating to the value of μ. The first, called the *Null Hypothesis* or H_0 for short, is that $\mu = \mu_0$, where μ_0 is some specified value. Typically, μ_0 will be the population mean for some related type of situation. For instance, if we are testing the response to a new treatment of some kind, μ_0 will usually be the mean response for the existing treatment. The Null Hypothesis is therefore the hypothesis of no change. The other hypothesis, called the *Alternative Hypothesis* or H_1, says that there has been some change, and may specify the nature of the change. We shall consider four types of Alternative Hypothesis: (i) $\mu \neq \mu_0$; (ii) $\mu > \mu_0$; (iii) $\mu < \mu_0$; (iv) $\mu = \mu_1$, where μ_1 is some specified value, not equal to μ_0. Just one of these types will be appropriate. Type (i) occurs when any change in the value of μ from μ_0 is of interest; cases (ii) and (iii) occur when only changes in a particular direction are worth detecting; type (iv) occurs when there is overwhelming evidence that the data can come only from one of two distributions: $N(\mu_0, \sigma^2)$ or $N(\mu_1, \sigma^2)$.

In the approach to hypothesis testing that we are going to adopt, it is assumed that the Null Hypothesis relates to a theory (perhaps suggested by previous experience) whose consistency with the experimental data we wish to examine. The implication is that we only want to decide against the Null Hypothesis if there are strong grounds for doing so. The hypotheses are not, therefore, treated symmetrically.

In the tyre manufacturer's problem, Illustration 5.1, the Null Hypothesis is that the new manufacturing method has made no difference to the mean life length, i.e. $H_0:\mu = 40$, where μ is the mean life length of the tyres made using the proposed method. Since there is only interest in increasing the life length, the Alternative Hypothesis is H_1: $\mu > 40$.

In order to see how we might decide whether the Null Hypothesis is acceptable, run program NORTEST. Choose the default values for the value of μ under H_0 and the variance (40 and 9 respectively). In the next option you are asked for the form of H_1; this is '$\mu >$', i.e. case 2. There are 10 tyres to be tested, so select the default value for the sample size. Since the Null and Alternative Hypotheses both relate to the value of the population mean, it seems sensible to base our decision between them on the value of $\bar{X}$, the sample mean. (More advanced texts provide theoretical justification for summarising the data in this way.) Recall that for normal data, the distribution of $\bar{X}$ is $N(\mu, \sigma^2/n)$. Select the 'Plot pdf of $\bar{X}$' option and choose $\mu = 40$. We see that if H_0 is true, the sample mean life length will probably lie between about 37 and 43. Suppose instead that H_1 is true; for instance, the mean life length may have increased to 42. Select the 'Plot pdf of $\bar{X}$' option again and choose this as the value of μ. The pdf of $\bar{X}$ when $\mu = 42$ is now superimposed. Notice that in this case, $\bar{X}$ will almost certainly lie between about 39 and 45. As we would expect, $\bar{X}$ will tend to be larger under H_1, but there is quite a considerable range of values of $\bar{X}$ which are reasonably consistent with μ being either 40 or 42, i.e. there will be no clear-cut way of deciding between H_0 and H_1.

If you try plotting the pdf of $\bar{X}$ for $\mu = 41$, you will find that there is an even wider range of values of $\bar{X}$ for which there is no clear-cut decision. On the other hand, if $\mu = 45$ (i.e. the new method of construction leads to a dramatic increase in the mean life length) there will almost certainly be little doubt that H_1 is true on the basis of the value of $\bar{X}$.

In general, values of $\bar{X}$ greater than 40 are going to give some support to H_1, rather than H_0, but how much bigger does $\bar{X}$ have to be before we opt for H_1? There are various approaches to answering this kind of problem. In this section we shall discuss the classical hypothesis-testing approach. In § 5.2 we shall consider the related 'significance' method, and some other approaches will be mentioned in § 5.5.

Significance level

In the hypothesis testing approach, it is assumed that primary importance is placed on avoiding the erroneous rejection of the Null Hypothesis. Specifically, we set in advance a value called the *significance level*. This is the probability of deciding for the Alternative Hypothesis, when really the Null Hypothesis is true. Significance levels chosen in practice are, naturally, quite small quantities; 0.05 (i.e. 1/20) is a popular value, but smaller values such as 0.01 and 0.001 are used when it is particularly important that the Null Hypothesis is not rejected wrongly.

In the tyre example, the manufacturer would wish to avoid making the change in the method of construction, with its associated costs and disruption, if there was really no increase in mean life length, i.e. he wants to limit the chance of opting for H_1 when really H_0 is true. We shall suppose that he sets the probability of erroneously deciding against H_0 at 0.1, i.e. the significance level is to be 0.1, or 10%. (We have chosen this high value to make the forthcoming graphical display clearer.)

Select the 'Display hypothesis test' option and the default value of 0.1 for the significance level. The resulting display is reproduced in figure 5.1. You can see that the part of the tail of the pdf of $\bar{X}$ under H_0 to the right of 41.22 is shaded. This value is chosen so that the area of the shaded region is 0.1, i.e. the probability of $\bar{X}$ taking a value in excess of 41.22 is 0.1, if H_0 is true. Note that 41.22 is the 90% point of the distribution of $\bar{X}$ under H_0, and may be found by using formula (2.5.8) and table 2 of Appendix 4. The flashing messages indicate how decisions are to be made on the basis of the value of $\bar{X}$. If $\bar{X}$ exceeds 41.22, we opt for H_1; otherwise we cannot reject H_0. This rule for deciding on the acceptability of H_0 satisfies the two conditions that we have already mentioned: (i) we only opt for H_1 if $\bar{X}$ is sufficiently large; (ii) if H_0 is true, the probability of erroneously opting for H_1 (i.e. $\bar{X} \geqslant 41.22$) is equal to 0.1. The values of $\bar{X}$ for which H_0 is rejected form the *critical region* of the test. (Notice that if we found that $\bar{X} = 41$, the above rule says that H_0 should not be rejected. This is not to say that the balance of evidence does not support H_1, which it patently does, but simply that the evidence for H_1 is not sufficiently strong that H_0 can be rejected at this significance level. For this reason, we have avoided saying that if $\bar{X} < 41.22$ then H_0 is 'accepted'. We shall discuss this point in more detail in § 5.5.) The rule that we have described for deciding between the two hypotheses is called a 'statistical test', or simply a 'test'.

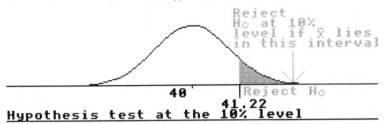

$H_0 : \mu = 40$ $H_1 : \mu > 40$ Sample size=10
Pdf of $\overline{X}$ under H_0

Reject H_0 at 10% level if $\overline{X}$ lies in this interval

40

41.22

Reject H_0

Hypothesis test at the 10% level

Figure 5.1 Finding the critical region of the test $H_0 : \mu = 40$ against $H_1 : \mu > 40$ for normally distributed data with $n = 10$, $\sigma^2 = 9$, using program NORTEST. The critical value, 41.22, is chosen so that the shaded area is equal to 0.1, the significance level.

Behaviour in repeated experiments

In order to see the test in action, stop the flashing message, choose the 'Single simulation' option and set the value of μ for the simulation equal to 40, the value under H_0. A value of $\overline{X}$ is simulated from the $N(40, 0.9)$ distribution (i.e. under H_0) and is plotted on the x axis of the graph. If this value is less than 41.22, the decision is not to reject H_0. This is the correct decision since we have set $\mu = 40$. Repeat this option a few times. You should find that H_0 is rejected only in a minority of cases.

We can check empirically on the percentage of times that H_0 is rejected by selecting the 'Continued simulation' option and again choosing $\mu = 40$. Initially, the percentage will wander around quite a bit as a result of random variation, but after a few hundred simulations it will settle down to values around 10%. Hence in any one experiment the probability of rejecting H_0 when it is true is 0.1, as we claimed.

It is natural to ask what would happen if the value of μ was greater than 40. This introduces the concept of the 'power' of the test, which we are going to discuss in some detail in § 5.3. For the moment, however, you can get a feel for what happens by selecting the 'Continued simulation' option again and a value for μ in excess of 40.

Varying the significance level

Now that we have seen one statistical test in operation, let us examine the effect of changing some of the assumptions. Select the 'Display hypothesis test' option and set the significance level to 0.05 (i.e. 5%). A second graph of the pdf is drawn, but this time the critical region is somewhat smaller, only consisting of the interval $\bar{X} \geqslant 41.56$. The reason is that the smaller significance level means that there is to be a smaller chance of erroneous rejection of H_0, i.e. the evidence against H_0, in terms of a 'large' value of $\bar{X}$, must be that much stronger before H_0 can be rejected.

Exercise
Choose significance levels of 0.01 (1%) and 0.2 (20%) and observe the effects on the critical region. Check that you understand why the regions vary in size in the way that they do.

Exercise
Suppose that the manufacturer decided to adopt the new method of manufacture if the sample mean life length exceeded 42.0. Use program NORTEST to find the significance level that he is effectively adopting. (You will need to use a certain amount of 'trial and error'.)

Varying the sample size
Let us now see the effect of changing the sample size. Select the 'Change the sample size' option and set the sample size to 20. Now select the 'Display hypothesis test' option, with a significance level equal to 0.05, and you will see that the pdf that is drawn is less spread out and more peaked than the previous ones. This is because the variance of the sample mean is equal to the variance of the underlying observations divided by the sample size, and so decreases as the sample size increases. An area of 0.05 under the right-hand tail of the pdf is drawn, as before, but since the pdf is more peaked the critical value is smaller, 41.10 rather than 41.56. The underlying reason is that, since there are more data, the sample mean does not have to exceed 40 by such a great amount before there is significant evidence against H_0 at the 5% level. More briefly, the greater the amount of data, the easier it is to arrive at firm conclusions. If you set the sample size to 100, you will find that the critical value is even closer to 40. On the other hand, if you set it to 4 the critical region is pushed back to the right, i.e. $\bar{X}$ must suggest a much more dramatic increase in life length before the evidence can be considered significant.

Different forms of Alternative Hypothesis
The final aspect of this example that we can change is the form of H_1. Firstly, though, redraw the diagram for a sample size of 10 and a significance level of 0.1. Now select the 'Change H_1' option and choose

case 1 for the form of H_1, i.e. $\mu \neq 40$. Select the 'Display hypothesis test' option with a significance level of 0.1. You will see that the critical region now consists of two intervals: $\bar{X} \leqslant 38.44$ and $\bar{X} \geqslant 41.56$. The total area of the two shaded regions is 0.1, so that the probability of rejecting H_0 when it is true is still 0.1, the significance level. You can confirm this empirically by selecting the 'Continued simulation' option with $\mu = 40$.

The reason for the change in the form of the critical region is that $\mu \neq 40$ under H_1 and so values of $\bar{X}$ either much greater than or much less than 40 tend to favour H_1. The critical values 38.44 and 41.56 are chosen so that each tail area is equal to 0.05, i.e. if H_0 is wrongly rejected, it is as likely that this is because $\bar{X}$ happens to be suspiciously small as it is that $\bar{X}$ happens to be suspiciously large.

The manufacturer might decide on this type of Alternative Hypothesis if the change in method of construction consisted of an increase in the amount of a certain substance in the composition of the tyres. If $H_0: \mu = 40$ can be rejected in favour of $H_1: \mu \neq 40$ at the 10% level, then there is significant evidence either that the additional substance leads to greater life length (if $\bar{X} > 40$) or to reduced life length (if $\bar{X} < 40$). The latter case might suggest a further experiment in which the proportion of the substance is reduced from its usual level.

Use the 'Change H_1' option to set H_1 to $\mu < 40$. Select the 'Display hypothesis test' option with a significance level equal to 0.1. This time you will find that the left-hand tail is shaded. Once again, the area of the shaded region is 0.1 in order to ensure that the significance level is 0.1. The critical region is $\bar{X} \leqslant 38.78$, i.e. there is significant evidence that the mean life length has dropped below 40 only if the sample mean life length has dropped to at least 38.78. The manufacturer might choose $\mu < 40$ as the Alternative Hypothesis if he wanted to develop a tyre which lasted for a shorter time and thereby, hopefully, generate more sales!

The final form of Alternative Hypothesis is $H_1: \mu = \mu_1$, where μ_1 is a specified value and $\mu_1 \neq \mu_0$, the value of μ under H_0. Suppose, for instance, that in the tyre example there was a specific claim that the new mode of construction led the mean life length to increase to 42. The manufacturer therefore wanted to decide whether there was significant evidence in favour of this claim rather than the Null Hypothesis of no change from the usual mean life length. In this case we have $H_0: \mu = 40$ and $H_1: \mu = 42$. In order to see the effect on the test, choose the 'Change H_1' option, case 4 for the form of H_1 and set $\mu = 42$ under H_1. Choose the 'Display hypothesis test' option with significance level 0.1. Compare the displayed diagram with the one for $H_0: \mu = 40$ against $H_1: \mu > 40$.

You will find that the critical regions are just the same. In this approach, therefore, the case $H_1 : \mu = 42$ leads to the same test as the case $H_1 : \mu > 40$. In fact, the same test would be obtained for $H_1 : \mu = \mu_1$ for any value of μ_1 greater than 40.

A note on terminology: hypotheses such as $\mu = 40$ or $\mu = 42$, in which the distribution is uniquely specified, are known as *simple hypotheses*. A hypothesis such as $\mu > 40$, in which more than one distribution is consistent with the hypothesis, is known as a *composite hypothesis*.

Exercise
Investigate what happens when μ_1 is less than 40.

Mathematical details
The formulae for the critical regions for the various forms of the Alternative Hypothesis are given in table 5.1.

Table 5.1   The critical regions for the α-level test of $H_0 : \mu = \mu_0$ against H_1, for various forms of H_1, where the data consist of a random sample $X_1, ..., X_n$ from the $N(\mu, \sigma^2)$ distribution, where σ^2 is known.

Form of Alternative Hypothesis	Critical region
(i) $H_1 : \mu \neq \mu_0$	Either $\bar{X} \leqslant \mu_0 - z_{1-\alpha/2}\,\sigma/n^{1/2}$ or $\quad \bar{X} \geqslant \mu_0 + z_{1-\alpha/2}\,\sigma/n^{1/2}$
(ii) $H_1 : \mu > \mu_0$	$\bar{X} \geqslant \mu_0 + z_{1-\alpha}\,\sigma/n^{1/2}$
(iii) $H_1 : \mu < \mu_0$	$\bar{X} \leqslant \mu_0 - z_{1-\alpha}\,\sigma/n^{1/2}$
(iv) $H_1 : \mu = \mu_1$	As for $H_1 : \mu > \mu_0$ if $\mu_1 > \mu_0$ As for $H_1 : \mu < \mu_0$ if $\mu_1 < \mu_0$

We shall indicate how the critical region is found for the case $H_1 : \mu > \mu_0$ only. Run program NORTEST with the default values of the parameters and H_1 of the form $\mu > \mu_0$ and choose the 'Display hypothesis test' option. Confirm that the critical region is of the form '$\bar{X} > a$', where a is chosen so that $P(\bar{X} \geqslant a) = \alpha$, the significance level of the test, the probability being evaluated assuming that H_0 is true. The distribution of $\bar{X}$ is $N(\mu_0, \sigma^2/n)$, so that a is the $100(1 - \alpha)\%$ point of the $N(\mu_0, \sigma^2/n)$ distribution. Hence, from (2.5.8),

$$a = \mu_0 + z_{1-\alpha}\,\sigma/n^{1/2}$$

where $z_{1-\alpha}$ is the $100(1 - \alpha)\%$ point of the $N(0, 1)$ distribution, found by using table 2 of Appendix 4.

Worked example
Use table 5.1 to find the critical region for the tyre example (where $H_0 : \mu = 40$, $H_1 : \mu > 40$, $n = 10$, $\sigma^2 = 9$) if the significance level is 1%.
  Answer. We see that $\mu_0 = 40$, $\alpha = 0.01$, $\sigma = 3$. From table 2, $z_{1-0.01} = z_{0.99} = 2.3263$. The critical region is therefore $\bar{X} \geqslant 40 + 3 \times 2.3263/\sqrt{10}$, i.e. $\bar{X} \geqslant 42.21$.

Exercise
A doctor is interested in the effect of bran in the diet. One part of her studies involves testing whether a greater proportion of bran leads to a change in weight gain in growing rats. She knows that for rats on the standard diet, the mean weight gain over a 30-day period is 10 g. She plans to study the effect of an increased proportion of bran on 16 rats in a 30-day experiment. In this case, H_0 is $\mu = 10$, where μ is the mean weight gain of rats on the bran-enhanced diet in the 30-day period, and H_1 is $\mu \neq 10$, since the doctor wishes to detect either an increase or a decrease. The distribution of weight gain under both the standard and experimental diets can be assumed to be normal with variance 300 g^2. If the sample mean weight gain of the 16 rats fed on the experimental diet is 19.1 g, can H_0 be rejected at (i) the 5% level; (ii) the 1% level; (iii) the 0.1% level? (Answer by using table 5.1 first and then check your results by running program NORTEST.)

Exercise
An anthropologist has discovered a prehistoric grave and has obtained five skulls, belonging to adult males of a certain type of early man. The standard theory relating to graves in the area of discovery states that the occupants belonged to type A, whilst the other possibility is that they belonged to type B. The two types differ in that the mean brain size for type A is 500 ml, whilst for type B it is 600 ml. Thus the Null Hypothesis is $\mu = 500$, i.e. no difference from the standard theory, and the Alternative Hypothesis is $\mu = 600$. The sample mean brain size of the five specimens is found to be 575 ml. For both types, the distribution of the brain size can be assumed to be normal with variance 15 000 ml^2. Can the standard theory, that the skulls belonged to type A, be rejected at the 5% level? (Answer first by using table 5.1 and verify your result using program NORTEST.) With the test you are using, what is the probability

that the standard theory will be rejected at the 5% level when really the skulls belonged to type B? (Use the 'Continued simulation' option of program NORTEST with $\mu = 600$ in order to obtain an estimate of this probability.)

)5.2 Significance

The hypothesis-testing approach that we have developed so far is very much concerned with making clear-cut decisions. Either the Null Hypothesis is rejected, or it is not. This implies that once the decision is made there is no need to undertake further similar experiments. In many situations this is too much of a simplification. Any one particular experiment may indicate, with some degree of strength, a certain conclusion, but this conclusion is not likely to be generally accepted until a number of independent similar experiments have been performed and have produced similar results. (Such is the case, for instance, in the sciences and in medical research.) In these circumstances, what is required is not a decision but a measure of the strength with which an experiment supports a given conclusion.

The measure that we shall use is called the *significance*. The basic set-up that we shall assume is the one we introduced in the last section. A random sample $X_1, X_2, ..., X_n$ is drawn from the $N(\mu, \sigma^2)$ distribution, so that the sample mean $\bar{X}$ has the $N(\mu, \sigma^2/n)$ distribution. The value of σ^2 is known and the value of μ is unknown. It is desired to test $H_0 : \mu = \mu_0$ against an Alternative, H_1. Suppose that the value of the sample mean in a particular experiment is $\bar{x}$. Then the significance of the outcome $\bar{x}$ is defined as

> the probability that, when H_0 is true, $\bar{X}$ takes a
> value as, or more, unfavourable to H_0 as the
> observed outcome $\bar{x}$.   (5.2.1)

This is quite a mouthful, so do not try to take in the full meaning at the moment. Basically, the significance is a measure of the strength with which the data accord with H_0; the smaller its value, the more strongly the data support H_1. Since it is a probability, its value must lie between 0 and 1.

We shall again use the tyre example of Illustration 5.1 to demonstrate the method. Recall that in this example, $\sigma^2 = 9$, $n = 10$ and $H_0 : \mu = 40$. We shall first consider the case $H_1 : \mu > 40$. Run program NORTEST and set the parameters in the usual way. Select the 'Evaluate significance for specified $\bar{x}$' option. The graph of the pdf of $\bar{X}$ under H_0 (the $N(40,0.9)$ distribution) is plotted. Suppose that the 10 tyres have been tested and their sample mean life length is 41. Enter the value 41 and you will see it indicated on the graph. Press the space bar and you will see the area under the pdf to the right of 41 being shaded. This is because it is values of $\bar{X}$ greater or equal to 41 that are 'as, or more, unfavourable to H_0' as the actual outcome, $\bar{x} = 41$. In other words, the observed sample mean of 41 lends some support to H_1; values of the sample mean in excess of 41 would have lent even more support to H_1, whilst values less than 41 would have been more favourable to H_0. The significance of the outcome $\bar{x} = 41$ is, according to (5.2.1), the probability of the event '$\bar{X} \geqslant 41$', assuming that $\mu = 40$. But this probability is just the area of the shaded region. On pressing the space bar to obtain the display shown in figure 5.2, you will see that this area is about 0.146.

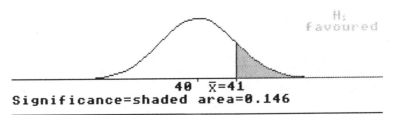

```
Ho:µ=40 H₁:µ>40 Sample size=10
           Pdf of x̄ under Ho
```

Figure 5.2   A display from program NORTEST after selecting the 'Evaluate the significance for specified $\bar{x}$' option and setting $\bar{x}$ to 41. The test is of $H_0 : \mu = 40$ against $H_1 : \mu > 40$, where the data are assumed to follow a normal distribution with $n = 10$, $\sigma^2 = 9$.

Suppose now that the sample mean life length of the experimental tyres had been 42. Select the 'Evaluate significance...' option and set $\bar{x} = 42$. This value is plotted and we can immediately see that such a value is much less in accord with H_0 than $\bar{x} = 41$. Press any key and the area to the right of 42, corresponding to values of $\bar{X}$ even less favourable to H_0, is shaded. Press any key to find that the significance is approximately 0.018. The smallness of this value indicates the strength of support for H_1 provided by the result $\bar{x} = 42$.

Now see what happens if $\bar{x} = 39$. Common sense suggests that such an outcome favours H_0. The high value of the significance indicated by the computer, about 0.85, confirms this.

P value

When reporting experimental results, the significance is usually denoted by P (or p). Thus, the statement '$P = 0.42$' would suggest that the results of the experiment were quite consistent with H_0, whilst the statement '$P = 0.02$' would suggest quite strong evidence in favour of H_1.

As is to be expected, there is a strong connection between 'significance' and 'significance level'. In order to see this, first evaluate the significance of the outcome $\bar{x} = 41.22$. You should find that the value of P is about 0.1. Now select the 'Display hypothesis test' option and set the significance level to 0.1. We see that the test at the 0.1 level is 'Reject H_0 if $\bar{X} \geqslant 41.22$'. Comparing the two diagrams, we see that any value of $\bar{x}$ which has significance less than or equal to 0.1 leads to the rejection of H_0 at the 0.1 level, whilst H_0 is not rejected if the significance of $\bar{x}$ exceeds 0.1. In general,

H_0 is rejected at significance level α if, and only if,
the significance is less than or equal to α.   (5.2.2)

Try evaluating the significance of the outcome 41.56 and comparing it with the 5% hypothesis test to confirm this result with $\alpha = 0.05$.

This relation between significance and significance level leads to the reporting of experimental results which indicate that H_0 can be rejected at the 5% level, say, as '$P < 0.05$' or 'the evidence against H_0 is significant at the 5% level'.

Varying the sample size

Let us now investigate the effect on the significance of a change in the sample size. First, recalculate the significance for $\bar{x} = 41$. Select the 'Change the sample size' option, set the sample size to 20 and calculate the significance for $\bar{x} = 41$. The value 41 appears, of course, at the same point on both graphs and the outcomes 'as, or more, unfavourable to H_0' are $\bar{X} \geqslant 41$ in both cases. The significance in the second case is smaller (about 0.068 rather than 0.146) since the pdf of $\bar{X}$ under H_0 is more peaked as a result of the larger sample size. This reflects the fact that a sample mean of 41 indicates much stronger evidence in favour of H_1 when the sample size is 20 than when it is 10. Setting the sample size to 80, you will find that the significance has dropped to about 0.001, i.e.

the outcome $\bar{x} = 41$ in this case indicates almost conclusive evidence in favour of H_1.

Other forms of Alternative Hypothesis

Suppose instead that the Alternative Hypothesis is $H_1 : \mu < 40$. Use the 'Change H_1' option to reset H_1 and the 'Change the sample size' option to reset the sample size to 10. Select the 'Evaluate significance...' option and set $\bar{x}$ to 41. The value 41 is indicated on the graph of the pdf of $\bar{X}$ under H_0 in the familiar way. On pressing any key, we discover the effect of changing H_1. The values of $\bar{X}$ that tend to favour H_1 more than the outcome $\bar{x} = 41$ are now the values less than 41. The significance is equal to the probability of this under H_0 (i.e. the area under the pdf to the left of 41) and is found by pressing any key to be about 0.854. The fact that the significance is so large reflects, of course, the conclusion that a sample mean of 41 provides precious little support for the hypothesis that $\mu < 40$.

Exercise

For the case $H_1 : \mu < 40$, find the significances of the outcomes $\bar{x} = 40$, 38.5 and 37. Confirm that the statement (5.2.2) continues to hold in this case.

The case $H_1 : \mu = \mu_1$ is easily dealt with. If $\mu_1 < \mu_0$, evaluate the significance as if H_1 were $\mu < \mu_0$, and if $\mu_1 > \mu_0$ evaluate the significance as if H_1 were $\mu > \mu_0$.

The final case, $H_1 : \mu \neq 40$, requires a little care. In program NORTEST, reset H_1 using the 'Change H_1' option and set the sample size to 10. Choose the 'Evaluate significance...' option and set $\bar{x}$ to 41. Before continuing, decide what you think are the values of $\bar{X}$ 'as, or more, unfavourable to H_0' than the outcome $\bar{x} = 41$ in this case. Once you have given this some thought, press the space bar. Not unexpectedly, the area corresponding to values of $\bar{X}$ greater than 41 is shaded. But since $H_1 : \mu \neq 40$, values of $\bar{X}$ much smaller than 40 are also unfavourable to H_0, and favourable to H_1. Press any key and note that the area to the left of 39 is shaded, as well as the area to the right of 41. The two shaded areas are equal since the pdf is symmetric. Press the space bar to evaluate the significance. The value obtained, about 0.292, is, of course, twice the significance of $\bar{x} = 41$ for $H_1 : \mu > 40$. (Roughly speaking, the more precisely we define what we are looking for, the more likely we are to find it; the hypothesis $\mu > 40$ is more precise than the hypothesis $\mu \neq 40$.)

Try evaluating the significance of the outcome $\bar{x} = 38.5$. Again, decide what you think the shaded region should be prior to drawing it on the screen.

Let us confirm that this 'two-tail' approach fits in with the idea of a significance level. Firstly, evaluate the significance at $\bar{x} = 41.56$. The combined area of the two tails is about 0.1. Now use the 'Display hypothesis test' option and set the significance level to 0.1. You will see that identical areas under the pdfs are shaded, indicating that the outcome $\bar{x} = 41.56$, having significance 0.1, is on the borderline of leading H_0 to be rejected at the 0.1 level. It follows that the test 'Reject H_0 if the significance is less than or equal to 0.1' has significance level 0.1, in accord with statement (5.2.2).

Mathematical details

The formulae for the significance for the various forms of the Alternative Hypothesis are summarised in table 5.2.

Table 5.2   The significance of the outcome $\bar{X} = \bar{x}$ when testing $H_0 : \mu = \mu_0$ against H_1 for various forms of H_1, where the data consist of the random sample $x_1, ..., x_n$ from the $N(\mu, \sigma^2)$ distribution and σ^2 is known.

Form of Alternative Hypothesis	Significance	
(i) $H_1 : \mu \neq \mu_0$	$2\Phi\left(\dfrac{\bar{x} - \mu_0}{\sigma/n^{1/2}}\right)$	if $\bar{x} < \mu_0$
	$2\left[1 - \Phi\left(\dfrac{\bar{x} - \mu_0}{\sigma/n^{1/2}}\right)\right]$	if $\bar{x} > \mu_0$
(ii) $H_1 : \mu > \mu_0$	$1 - \Phi\left(\dfrac{\bar{x} - \mu_0}{\sigma/n^{1/2}}\right)$	
(iii) $H_1 : \mu < \mu_0$	$\Phi\left(\dfrac{\bar{x} - \mu_0}{\sigma/n^{1/2}}\right)$	
(iv) $H_1 : \mu = \mu_1$	As for $H_1 : \mu > \mu_0$	if $\mu_1 > \mu_0$
	As for $H_1 : \mu < \mu_0$	if $\mu_1 < \mu_0$

We shall now indicate how the formula for the case $H_1 : \mu < \mu_0$ is obtained. Use program NORTEST with the default values of the parameters and this form of H_1 to confirm that the significance is equal to $P(\bar{X} \leqslant \bar{x})$, the probability being evaluated assuming that $\mu = \mu_0$. Now, since the distribution of $\bar{X}$ is then $N(\mu_0, \sigma^2/n)$,

$$P(\bar{X} \leqslant \bar{x}) = \Phi\left(\frac{\bar{x} - \mu_0}{\sigma/n^{1/2}}\right)$$

where Φ is the distribution function of the standard normal distribution (see formula (2.5.5)).

The other formulae in table 5.2 may be confirmed in a similar way.

Worked example
Suppose that in a hypothesis test, $n = 20$, $\sigma^2 = 8$, $H_0 : \mu = 30$, $H_1 : \mu < 30$ and the value of $\bar{x}$ is found to be 28.5. Evaluate the significance.

Answer. From table 5.2, the significance is equal to

$$\Phi\left(\frac{28.5-30}{(8/20)^{1/2}}\right) = \Phi(-2.372).$$

From table 1 (Appendix 4), or program DISTN, $\Phi(-2.372) = 0.0089$.

Exercise
In the 'skulls' exercise, page 103, what is the significance of the observed value of the sample mean, 575 ml? (Use the appropriate formula in table 5.2 and check your answer by using program NORTEST.)

Exercise
In the 'bran' exercise, page 103, evaluate the significance of the given sample mean weight gain, without and with the help of program NORTEST. How can this value be used to answer the questions concerning the level at which H_0 can be rejected?

⟩5.3 Power

We have seen that a hypothesis test is a rule for deciding whether or not to reject the Null Hypothesis H_0 on the basis of the data. We hope, of course, that the decision is correct as often as possible, but, because of the random nature of the data, there is always a chance of making a mistake. The kind of mistake that we have concentrated on so far is the one that occurs when the test wrongly rejects H_0; this is conventionally called the *Type I error*. The significance level is chosen so as to limit the probability of the Type I error occurring. The other kind of error occurs when the test does not reject H_0, even though the Alternative Hypothesis

H₁ is true. This is called a *Type II error*. In this section we shall consider
the probability of this kind of error, introducing the new ideas in the context of the tyre example, Illustration 5.1.

Recall that in this example, the life length of the new type of tyre (in
1000s of miles) is assumed to follow the $N(\mu, 9)$ distribution. The Null
Hypothesis is $H_0 : \mu = 40$ and the significance level is set at 0.1. We shall
first consider the simple Alternative Hypothesis, $H_1 : \mu = 42$, a case that
we considered towards the end of § 5.1.

Run program POWER, setting the value of μ under H_0 to 40, σ^2 to
9, the form of H_1 to 4, μ_1 to 42, the sample size to 10 and the significance
level to 0.1. Select the 'Evaluate power at specified value of μ' option.
A plot of the pdf of the sample mean $\bar{X}$ under H_0 is drawn and the
critical region is indicated. Suppose now that H_1 is correct; what is the
probability that H_0 is rejected by the test? Type the value 42 in response
to the 'value of μ' query on the computer. You will see the pdf of $\bar{X}$ when
$\mu = 42$ superimposed on the graph. The display is reproduced in figure
5.3. (Since, in general, $\bar{X}$ has the $N(\mu, \sigma^2/n)$ distribution , the displayed
pdf is of the $N(42,0.9)$ distribution.) The shaded area corresponds to
that part of the pdf within the critical region. In the usual way for pdfs,
this area is equal to the probability that $\bar{X}$ lies within the critical region
when $\mu = 42$, i.e. it is the probability that H_0 is correctly rejected. This
probability is called the *power* of the test. The computer tells us that in
this case the power is approximately 0.796.

We see, therefore, that there is nearly a 20% chance that the Null
Hypothesis will not be rejected even if the mean life length for tyres made

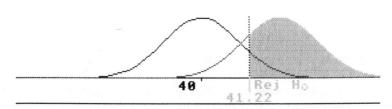

Figure 5.3   Finding the power at $\mu = 42$ of the test of $H_0:\mu = 40$ against
$H_1:\mu = 42$, for normally distributed data, $n = 10$, $\sigma^2 = 9$ and significance level 0.1,
using program POWER. The left-hand curve is of $N(40,0.9)$ (the distribution
of $\bar{X}$ under H_0) and the right-hand curve is of $N(42,0.9)$ (the distribution of $\bar{X}$
under H_1). The power is equal to the shaded area.

by the new method has increased to the claimed value of 42. The manufacturer might well be concerned that there is such a large probability of missing a promising development in his product. One possibility is to test more tyres. Suppose, for instance, that 20 instead of 10 tyres are produced by the new method; what is the effect on the power? Choose the 'Change the sample size' option, set the new sample size to 20 and select the 'Evaluate power ...' option again. The pdf of $\overline{X}$ is plotted and is seen to be rather more peaked than the previous pdfs since the variance is now 9/20 rather than 9/10. Set $\mu = 42$ and the pdf under H_1 is plotted. The power, or probability that H_0 is rejected when H_1 is true, is again equal to the shaded area. Since the two pdfs are more peaked, the power has increased to about 0.955. Thus, the effect of conducting a more extensive experiment would be to reduce the probability of erroneous failure to reject H_0 from about 0.2 to about 0.045.

Exercise
Investigate the effect on the power of increasing the sample size further. Roughly how many tyres need to be tested if the probability of rejecting H_0 at the 10% level when H_1 is true is to be 0.99?

However the sample size is changed in the above example, the probability of erroneous rejection of H_0 is always 0.1. Suppose that the manufacturer is concerned that this value is too high. How would decreasing it affect the power? In other words, how would reducing the probability of the Type I error affect the probability of the Type II error? Reset the sample size to 10 by using the 'Change the sample size' option and re-evaluate the power at 42. Select the 'Change the significance level' option and set the new significance level to 0.05. This corresponds to the manufacturer only being prepared to accept a 1 in 20 chance of wrongly deciding that the mean life length has increased to 42. Select the 'Evaluate power ...' option and set $\mu = 42$. The critical value has increased (so that the right-hand tail area under H_0 is reduced to 0.05) and the power (i.e. the shaded area) is consequently reduced to about 0.678. In round figures, the effect of halving the probability of the Type I error is to increase the probability of the Type II error from 0.2 to 0.3. So there is a 'cost' in reducing the significance level in terms of also reducing the power of the test.

Exercise
Suppose that the manufacturer decided to accept a 1 in 40 chance of

making either type of error (i.e. deciding that there had been an improvement in mean life length when there had not, or vice versa). Approximately how many experimental tyres must be made and tested? (An iterative approach is required.)

Exercise
Repeat the above exercise with the Alternative Hypothesis $\mu = 38$.

Power function
We have introduced the idea of the power of a test when the Alternative Hypothesis is simple (i.e. of the form $\mu = \mu_1$). Now let us consider the meaning of 'power' when the Alternative is composite. In the tyre example, for instance, suppose that the Alternative Hypothesis is $\mu > 40$, i.e. the manufacturer wishes to detect any increase in the mean life length. Restart program POWER, set $H_0 : \mu = 40$, $\sigma^2 = 9$, $H_1 : \mu > 40$, sample size 10 and significance level 0.1. Select the 'Evaluate power ...' option. The pdf of $\bar{X}$ is plotted and we are asked for the value of μ. This time, there is no single value of μ that presents itself; a Type II error will occur if the test does not reject H_0 for any value of μ greater than 40.

We have to start somewhere, so let us set $\mu = 42$, as before. The power, as we have already seen, is about 0.796. Since the choice of value for μ was arbitrary, however, this cannot by itself be regarded as a satisfactory summary of the properties of the test. What we need to know is the probability of making a Type II error for any value of μ greater than 40. Let us therefore make a plot of these probabilities by selecting the 'Plot the power' option. The power we have just evaluated is plotted, using a '*' for emphasis. Select the 'Evaluate power. . .' option and set $\mu = 41$. This time the power is only about 0.41; not unexpectedly, a smaller increase in the mean life length from the Null Hypothesis value leads to a smaller chance of the Null Hypothesis being rejected. This value is also plotted. Use the 'Evaluate power. . .' option to plot the power at 40.5, 43.0 and 44.0.

The plotted points appear to lie on a curve which increases fairly steeply from 40.5 to about 42 and less steeply thereafter. In order to see the curve in more detail, select the 'Plot the complete power function' option. This curve, the *power function*, is denoted by $\pi(\mu)$, where

$$\pi(\mu) = P(H_0 \text{ rejected when population mean} = \mu).$$

Since $\pi(\mu)$ is a probability, its value must lie between 0 and 1 inclusive for all values of μ. It is small for $\mu < 40$, since if the new process were to lead to a reduction in the mean life length, the probability of $\bar{X}$ being in the critical region (suggesting an increase in the value of the mean) would be small. On the other hand, for values of μ larger than about 44, the power function is very nearly 1 since such a large increase in the mean life length is almost certain to be detected. Now use the 'Evaluate power ...' option to evaluate the power function at 40. The diagram indicates that $\pi(40)$ is equal to the significance level, 0.1. This can also be confirmed by recalling the definitions of significance level and power function.

The power function therefore summarises the characteristics of a test. By studying the power function, the manufacturer can decide whether the proposed experiment is sensitive enough to have a good probability of detecting 'worthwhile' increases in the mean tyre life length.

He can also compare power functions of two or more tests. For instance, suppose that 20 tyres were used in the trial. Reset the sample size using the 'Change the sample size' option and plot the complete power function. The new power function is superimposed on the graph, in a different shade or colour. It has a similar shape to the previous one, but increases more steeply to 1 for $\mu > 40$. As we would expect, there is a greater chance of detecting increases in the mean life length when the sample size is increased. For a more dramatic demonstration of this effect, plot the power function for a sample size of 1000. The new function approximates a 'step' function, taking the value 0 for $\mu < 40$ and 1 for $\mu > 40$. This means that the test is almost 'perfect' in the sense that it hardly ever rejects H_0 when $\mu < 40$ and almost certainly rejects H_0 when $\mu > 40$. The price of such 'perfection' is a very large experiment.

Exercise

The manufacturer wishes the 10% significance test of $H_0 : \mu = 40$ against $H_1 : \mu > 40$ to meet the following two criteria: the probabilities of failing to reject H_0 when $\mu = 41.5$, $\mu = 42.5$ are to be 0.1 and 0.01 respectively. What is the smallest sample size required to achieve these objectives and which one turns out to be the more restrictive? (Some experimentation is required.)

Exercise

Plot the power function for the 10% significance test and the 1% significance test (using the 'Change the significance level' option), with sample size 10 and $H_1 : \mu > 40$ in each case. Explain in qualitative terms the reason for the difference in shape of the two curves.

Exercise
How would you expect the power function to look if H_1 were $\mu < 40$? Restart program POWER and check whether you are correct. Investigate the effects of variations in the sample size.

The last variant of the tyre example that we shall consider in this section is the two-sided Alternative Hypothesis, $\mu \neq 40$. Restart program POWER, set $H_0 : \mu = 40$, $\sigma^2 = 9$, $H_1 : \mu \neq 40$, sample size 10 and significance level 0.1. Choose the 'Evaluate power ...' option and confirm that the test is now two sided, i.e. H_0 is rejected if $\bar{X} \leqslant 38.44$ or $\bar{X} \geqslant 41.56$. Set μ equal to 42. The pdf of $\bar{X}$ for this value of μ is plotted and, as usual, the power (about 0.679) is equal to the shaded area. Plot this value using the 'Plot the power' option. What is not clear from the first diagram is that there are contributions to the power from both parts of the critical region. To make this more apparent, evaluate the power at 40.3; both tails of the pdf under H_1 can now be seen to be shaded.

Now evaluate the power for $\mu = 38$. Instead of declining further, as was the case for $H_1 : \mu > 40$, the power has risen to about 0.679. This is what we would expect, since the test for the case $H_1 : \mu \neq 40$ is supposed to detect any change of μ from 40, not just increases. The symmetry of the normal curves guarantees that the power at 38 is equal to the power at 42.

Plot the complete power function. The shape is different from the previous functions that we have plotted; values of μ both much greater than and much less than 40 lead to high probabilities of H_0 being rejected. It is again of interest to investigate the effect of changing the sample size. Select the 'Change the sample size' option, set the sample size to 20 and plot the complete power function. The new function dips more sharply to the value 0.1 at $\mu = 40$, indicating a greater chance of rejecting H_0 when $\mu \neq 40$. Repeat this with a sample size of 1000 in order to obtain a near 'perfect' test, in the sense already described.

Mathematical details
The formulae for the power for the various forms of Alternative Hypothesis are summarised in table 5.3.

As an example of how these formulae are derived, we shall consider case (ii), $H_1 : \mu > \mu_0$. Recall that for this case the level-α test is to reject H_0 if

$$\bar{X} \geqslant \mu_0 + (\sigma/n^{1/2})z_{1-\alpha}.$$

Table 5.3 The power at μ for the α-level test of $H_0 : \mu = \mu_0$ against H_1, for various forms of H_1, where the data consist of the random sample $X_1, ..., X_n$ from the $N(\mu, \sigma^2)$ distribution, where σ^2 is known.

Form of Alternative Hypothesis	Power
(i) $H_1 : \mu \neq \mu_0$	$\pi(\mu) = \Phi\left(\dfrac{\mu_0 - \mu}{\sigma/n^{1/2}} - z_{1-\alpha/2}\right)$ $+ 1 - \Phi\left(\dfrac{\mu_0 - \mu}{\sigma/n^{1/2}} + z_{1-\alpha/2}\right)$
(ii) $H_1 : \mu > \mu_0$	$\pi(\mu) = 1 - \Phi\left(\dfrac{\mu_0 - \mu}{\sigma/n^{1/2}} + z_{1-\alpha}\right)$
(iii) $H_1 : \mu < \mu_0$	$\pi(\mu) = \Phi\left(\dfrac{\mu_0 - \mu}{\sigma/n^{1/2}} - z_{1-\alpha}\right)$
(iv) $H_1 : \mu = \mu_1$	$\pi(\mu_1)$ is as for $H_1 : \mu > \mu_0$ if $\mu_1 > \mu_0$ and as for $H_1 : \mu < \mu_0$ otherwise

The power at μ is the probability that this inequality is satisfied when the mean is equal to μ, i.e the distribution of $\bar{X}$ is $N(\mu, \sigma^2/n)$. But from (2.5.6),

$$P(\bar{X} > \mu_0 + (\sigma/n^{1/2})z_{1-\alpha}) = 1 - \Phi([\mu_0 + (\sigma/n^{1/2})z_{1-\alpha} - \mu]/(\sigma/n^{1/2}))$$

$$= 1 - \Phi\left(\frac{\mu_0 - \mu}{\sigma/n^{1/2}} + z_{1-\alpha}\right).$$

Worked example
Evaluate the power at $\mu = 42$ in the tyre example (where $H_0 : \mu = 40$, $H_1 : \mu > 40$, $n = 10$, $\sigma^2 = 9$ and $\alpha = 0.1$).
   Answer.

$$\pi(42) = 1 - \Phi\left(\frac{40 - 42}{3/\sqrt{10}} + z_{1-0.1}\right) = 1 - \Phi(-2.108 + 1.2816)$$

$$= 1 - \Phi(-0.826) = 1 - 0.2044 = 0.7956.$$

Exercise
Recall the 'skulls' exercise, page 103. Evaluate the power of the test. What would the power be if the significance level were 10%? Check your answers by using program POWER.

Exercise

For the 'bran' exercise, page 103, what is the probability of detecting $\mu = 15$ if the significance level is 5%? Verify your answer by using program POWER and use this program to compare the power functions for the tests with 5%, 1% and 0.1% significance levels. Explain why the functions do not intersect.

⟩5.4   Discrete distributions

So far in this chapter we have concentrated on data following the normal distribution. Most of the ideas that we have introduced extend in a natural way to other situations involving continuous distributions, although the application of the ideas is usually not so easy. We shall consider some examples in the next chapter. For discrete distributions, however, while the underlying principles remain the same, a number of modifications are required when evaluating quantities of interest. We shall use the binomial distribution for the purposes of demonstration, but the ideas apply equally to other discrete distributions with a single unknown parameter, such as the Poisson distribution.

Significance

We shall investigate the evaluation of the significance of binomially distributed random variables in the context of the following example.

Illustration 5.2

A market research study on the relative popularity of two similar-looking soft drinks, 'Kepsi' and 'Poke', is being undertaken. A random sample of 10 people are given a glass of each and asked which they prefer. (They must choose one or the other.) The manufacturers are interested in the value p, the proportion of the whole population preferring Kepsi. In particular, they want to know whether the two drinks are equally popular, i.e. if $p = 0.5$. We shall refer to the number in the sample preferring Kepsi as X.

The random variable X has the $B(10, p)$ distribution since the probability of each of the 10 people preferring Kepsi is p. (Recall the conditions for the occurrence of the binomial distribution described in § 2.4.) The Null Hypothesis is $p = 0.5$.

Run program BINTEST and set $H_0 : p = 0.5$ and $n = 10$. We shall consider the Alternative Hypothesis, $H_1 : p > 0.5$, first. Type '2' for the form of H_1 and select the 'Evaluate significance' option. A diagram of the $B(10,0.5)$ distribution, the distribution under H_0, is drawn. (A histogram is used rather than a barchart since we shall be considering the normal approximation later on.)

Suppose that in the study, 8 people prefer Kepsi, i.e. $X = 8$. Select this as the value of x and note that it is indicated on the graph. This outcome lends some support to H_1, but how much? According to definition (5.2.1), with $\bar{X}$ replaced by X, the significance of the outcome x is

> the probability that, when H_0 is true, X takes a
> value as, or more, unfavourable to H_0 as the
> observed outcome x.   (5.4.1)

Press the space bar and you will see that the part of the histogram corresponding to $X = 8, 9$ and 10 is shaded. This is because the outcomes 9 and 10 are more unfavourable to H_0, suggesting as they do that H_0 is wrong even more strongly than the outcome 8. The shaded area is therefore equal to the probability that $X \geqslant 8$, assuming that H_0 is true, i.e. it is equal to the significance. (Note that since the width of each rectangle is 1, the area of each is equal to the corresponding probability.) Press the space bar to evaluate the significance. The value is about 0.055, i.e. the evidence against H_0 just fails to be significant at the 5% level.

Exercise
Select the 'Binomial probabilities' option. How could the significance of the outcome $X = 8$ have been evaluated using the displayed probabilities? Use the probabilities to evaluate the significance of the outcome $X = 7$. Check your answer by using the 'Evaluate significance ...' option.

Exercise
Suppose that instead, $H_1 : p < 0.5$. Select the 'Binomial probabilities' option and evaluate the significance of the outcome $X = 1$. Check your answer by restarting the program, resetting the parameters appropriately and selecting the 'Evaluate significance' option.

Now let us consider the case $H_1 : p \neq 0.5$. Recall that for normal data we had to be careful to remember that both tails contributed to the significance when H_1 was two sided. The same applies here. Restart program BINTEST, set $H_0 : p = 0.5$, $n = 10$ and $H_1 : p \neq 0.5$. Select the 'Evaluate significance' option and set $x = 8$. The outcome $X = 8$ is indicated and, on pressing the space bar, the appropriate part of the right-hand tail is shaded. If the space bar is pressed again the corresponding part of the left-hand tail is shaded, since the values 0, 1 and 2 are as, or more, unfavourable to H_0 as the outcome 8. The significance is the total shaded area which, with one more press of the space bar, is evaluated to be about 11%. Notice that this is twice the value we found for the significance for the one-sided Alternative Hypothesis, $H_1 : p > 0.5$.

When the hypotheses are $H_0 : p = p_0$ and $H_1 : p \neq p_0$, for $p_0 \neq 0.5$, so that the binomial distribution is not symmetric under H_0, then the evaluation of the significance becomes more troublesome. The interested reader may investigate it further by using program BINTEST. The main point to note is that the area that is displayed second is as large as possible, subject to not exceeding the area displayed first.

Significance—normal approximation
The evaluation of binomial probabilities can be greatly facilitated by making use of the normal approximation. Recall from § 2.7 that this states that if $np(1 - p)$ is 'large' (> 8, say) then the $B(n,p)$ distribution can be approximated by the $N(np,np(1 - p))$ distribution, in a sense to be demonstrated. Rerun program BINTEST with $H_0 : p = 0.5$, $n = 20$ and $H_1 : p > 0.5$. Select the 'Evaluate significance' option and set $x = 13$. Press the space bar twice in order to evaluate the significance; then select the 'normal approximation' option. You will see that the pdf of a normal distribution is superimposed. The mean and variance of this distribution are equal to the mean and variance of the binomial distribution, i.e. $20 \times 0.5 = 10$ and $20 \times 0.5 \times (1 - 0.5) = 5$ respectively. It is apparent that the significance (i.e. the shaded area) is approximately equal to the area under the normal curve to the right of 12.5. The reduction of '13' to '12.5' occurs because the bars of the histogram spread 0.5 to each side of the integers. The approximate significance can therefore be easily evaluated by the use of normal tables. Mathematical details of this and related cases are given at the end of the section.

Exercise
Suppose instead that the Alternative Hypothesis were $H_1 : p < 0.5$ and the observed value of x were 6. Use program BINTEST to help you decide the area under the normal curve that would best approximate the significance.

Exercise
Repeat the previous exercise with $H_1 : p \neq 0.5$.

Hypothesis testing

We saw in § 5.2 that the Null Hypothesis can be rejected at significance level α if the significance of the observed outcome is less than or equal to α. Recall Illustration 5.2, where $H_0 : p = 0.5$, $H_1 : p > 0.5$ and $n = 10$, and suppose that we want to set up a hypothesis test with significance level 10%. Restart program BINTEST with these values and choose the 'Evaluate significance' option. The critical region of the test (i.e. the values of X for which the Null Hypothesis is to be rejected) consists of those values leading to a significance of less than or equal to 0.1. Before reading on, try a number of values of X to determine the critical region in this case.

You should find that the critical region consists of the values 8, 9 and 10, with significances of about $0.055, 0.011$ and 0.001 respectively. In fact the 'logical order' for choosing values for X is 10, 9, 8, 7, ..., i.e. starting with the value most favourable to H_1 and then moving to steadily less favourable values until the significance exceeds the significance level. If you now select the 'Display hypothesis test' option and set the significance level to 0.1, you will see that this is how the computer 'thinks out' the answer. Notice that the significance level of the test is not equal to the probability of the Null Hypothesis being rejected when it is correct, since this probability is equal to about 0.055, rather than 0.1. This is unfortunate, since our test is more stringent than we wished, but is a necessary consequence of dealing with a discrete distribution.

(It should be noted that there is a mathematical device called a 'randomised test' which enables the significance level to be attained exactly. In our example, this would take the form 'Reject H_0 if $X = 8, 9$ or 10; reject H_0 with probability 0.39 if $X = 7$; otherwise do not reject H_0'. This test has significance level 0.1 and is the 'best' in a well defined sense. However, the idea of deciding randomly whether or not to reject H_0 does not appeal to most applied statisticians, for understandable reasons.)

Exercise

Find the 5% significance test in the above example.

Exercise

It is believed that the probability p of obtaining a '6' on a certain die is greater than 1/6, and it is required to test $p = 1/6$ against the Alternative $p > 1/6$ at the 0.1% significance level on the basis of the number of 6's obtained in 50 throws. Find the critical region for the test.

Exercise
Suppose that X has the $B(15, p)$ distribution and it is required to test $H_0 : p = 0.4$ against $H_1 : p < 0.4$ at the 5% level. Before using the 'Display hypothesis test' option, select the 'Binomial probabilities' option and decide what the critical region will be. Use the 'Display hypothesis test' option to check your answer.

When the Alternative Hypothesis is two sided (i.e. $H_1 : p \neq p_0$), the hypothesis test again has a critical region equal to those values of X for which the significance is less than or equal to the significance level α. A reasonable approximation is obtained, if the distribution under H_0 is approximately symmetric, by making the critical region equal to the union of the critical regions of the two one-sided level-$\alpha/2$ tests (i.e. the tests with $H_1 : p < p_0$ and $H_1 : p > p_0$). Program BINTEST does not deal directly with this case.

Hypothesis testing—normal approximation

The normal approximation can sometimes be used to help find the critical region. If X has the $B(n, p)$ distribution and it is required to test $H_0 : p = p_0$, then the normal approximation will often be appropriate if $np_0(1 - p_0)$ is greater than about 8.

For example, rerun program BINTEST with the parameters $p_0 = 0.5$, $n = 20$ and $H_1 : p > 0.5$. (We are violating the '$np_0(1 - p_0) > 8$' rule in order to obtain a clearer diagram.) Select the 'Display hypothesis test' option, set the significance level equal to 0.1 and find the critical region of the test. Select the 'normal approximation' option and observe that the shaded area is approximately equal to the area to the right of 13.5 under the normal curve. In order to find the critical region without having to evaluate the binomial probabilities, we could find the value c, say, such that the area under the normal curve to the right of c is 0.1. In fact, c is approximately 12.9 in this case. Since we are approximating the binomial distribution, only values of the form 'integer plus 0.5' are allowed. The significance level is to be no greater than 0.1, so we must increase c to the next allowed value, 13.5, corresponding to the critical region $X \geqslant 14$. When the value of X is known, you can avoid these manipulations by making use of the connection between hypothesis tests and significance already described.

Exercise
Suppose that we want to find the 5% significance level test for

$H_0 : p = 0.6$ against $H_1 : p < 0.6$ when $n = 50$. The area under the pdf of $N(30, 12)$ to the left of 24.3 is approximately 0.05. What critical region does this suggest? Use program BINTEST to check your answer.

Power

Recall from § 5.3 that the power of a test of two simple hypotheses is the probability that H_0 is rejected when H_1 is true. Let us examine how the power is evaluated when the data have the binomial distribution by considering the case $H_0 : p = 0.5$, $n = 15$ and $H_1 : p = 0.7$. Restart program BINTEST, set the parameters and select the 'Display hypothesis test' option. Setting the significance level equal to 0.1, we find that the critical region consists of values of X greater than or equal to 11.

Select the 'Evaluate the power' option. The diagram of the distribution under H_0 is redrawn in outline form and the boundary of the critical region is indicated by a vertical line. On setting $p = 0.7$, the $B(15, 0.7)$ distribution is superimposed. The lighter shaded (or coloured) part of this distribution corresponds to the acceptance region and the darker shaded part corresponds to the critical region. The power (i.e. the probability that H_0 is rejected when $p = 0.7$) is the area of the darker shaded region, and is approximately equal to 0.515. Choose the 'Binomial probabilities' option to check that you know how this value is calculated.

Exercise

The random variable X has the $B(12, p)$ distribution and it is desired to test $H_0 : p = 0.7$ against $H_1 : p = 0.3$. The test 'Reject H_0 when $X \leqslant 4$' is used. By making use of the 'Binomial probabilities' option of program BINTEST with appropriate parameter values, evaluate the significance level and power of the test. Check your answers by making use of the 'Display the hypothesis test' and 'Evaluate the power' options.

Exercise

The random variable X has the $B(25, p)$ distribution and it is required to test $H_0 : p = 0.6$ against $H_1 : p < 0.6$. Use BINTEST to help plot on graph paper the power function of the 5% significance test. (Recall that the power function $\pi(p)$ is equal to the power at p for all values of p.)

Exercise

The success rate of the standard treatment for a certain disease is 20%. A new treatment has been proposed and a clinical trial is to be con-

Table 5.4 The significance of the outcome $X = x$, where X has the $B(n,p)$ distribution, when testing $H_0 : p = p_0$ against H_1, for various forms of H_1 $P(k)$ denotes $P(X = k)$ when $p = p_0$. The normal approximation is applicable when $np_0(1 - p_0)$ is 'large'—greater than 8, say.

Form of H_1	Exact significance	Significance using normal approximation
(i) $H_1 : p \neq p_0$	—	$2\left[1 - \Phi\left(\dfrac{\|x - np_0\| - 0.5}{[np_0(1 - p_0)]^{1/2}} \right) \right]$
(ii) $H_1 : p > p_0$	$\displaystyle\sum_{k=x}^{n} P(k)$	$1 - \Phi\left(\dfrac{x - 0.5 - np_0}{[np_0(1 - p_0)]^{1/2}} \right)$
(iii) $H_1 : p < p_0$	$\displaystyle\sum_{k=0}^{x} P(k)$	$\Phi\left(\dfrac{x + 0.5 - np_0}{[np_0(1 - p_0)]^{1/2}} \right)$
(iv) $H_1 : p = p_1$		As for $H_1 : p > p_0$ if $p_1 > p_0$ As for $H_1 : p < p_0$ if $p_1 < p_0$

ducted. How many patients should enter the trial if a 5% significance level is to be used and the probability of rejecting $H_0 : p = 0.2$ in favour of $H_1 : p > 0.2$ when $p = 0.35$ is to be about 0.8 (Hint: try different values of n until a 5% test with the correct power is found.)

Mathematical details
Formulae for the significance of the binomial outcome x are given in table 5.4.

Table 5.5 The approximate critical region for the α-level test of $H_0 : p = p_0$ against various forms of H_1, when X has the $B(n,p)$ distribution. The function ROUND converts the value in brackets to the nearest integer.

Form of H_1	Critical region (using normal approximation)
(i) $H_1 : p \neq p_0$	Either $X > \text{ROUND}\{ np_0 + z_{1-\alpha/2}[np_0(1 - p_0)]^{1/2} \}$ or $\quad X < \text{ROUND}\{ np_0 - z_{1-\alpha/2}[np_0(1 - p_0)]^{1/2} \}$
(ii) $H_1 : p > p_0$	$X > \text{ROUND}\{ np_0 + z_{1-\alpha}[np_0(1 - p_0)]^{1/2} \}$
(iii) $H_1 : p < p_0$	$X < \text{ROUND}\{ np_0 - z_{1-\alpha}[np_0(1 - p_0)]^{1/2} \}$
(iv) $H_1 : p = p_1$	As for $H_1 : p > p_0$ if $p_1 > p_0$, otherwise as for $H_1 : p < p_0$

Approximate formulae for the critical regions for the various forms of Alternative Hypothesis are given in table 5.5.

Worked example
Suppose that the random sample in Illustration 5.2 consists of 40 people (i.e. $n = 40$) and the hypotheses are $H_0 : p = 0.5$ and $H_1 : p \neq 0.5$. (i) Find the significance if 26 people prefer Kepsi. (ii) Find the critical region of the test at the 5% level.

Answer.(i) From table 5.4 we see that the approximate significance of the outcome $x = 26$ is

$$2\left(1 - \Phi\left(\frac{|26 - 40 \times 0.5| - 0.5}{[40 \times 0.5 \times (1 - 0.5)]^{1/2}}\right)\right) = 2(1 - \Phi(1.739)) = 2(1 - 0.959)$$

$$= 0.082.$$

(ii) We use table 5.5. The right-hand side of the first inequality is evaluated as follows:

$$\text{ROUND}\{40 \times 0.5 + [40 \times 0.5(1 - 0.5)]^{1/2} z_{1 - 0.05/2}\}$$
$$= \text{ROUND}(20 + \sqrt{10} \times 1.96) = \text{ROUND}(26.2) = 26.$$

The right-hand side of the second inequality is

$$\text{ROUND}\{40 \times 0.5 - [40 \times 0.5(1 - 0.5)]^{1/2} z_{1 - 0.05/2}\}$$
$$= \text{ROUND}(13.8) = 14.$$

The critical region therefore consists of values of X less than 14 or greater than 26. Use the 'Evaluate significance' option of program BINTEST with $X = 27$ and $X = 26$ to confirm that this is the test with largest significance less than 0.05.

Exercise
In a particular large prison, 60% of all inmates serving sentences of between 1 and 5 years return to prison within 3 years of release. A random sample of 100 such prisoners has been selected for a special programme of activities while they are in prison. It is required to evaluate the effectiveness of the programme by seeing how many of them return to prison within 3 years of release. (i) Find the critical region, if the significance level is to be 5%. (ii) What would the significance be if it were subsequently found that 48 of the 100 inmates returned to prison within 3 years? (Use a one-sided Alternative Hypothesis.)

⟩**5.5   The use of statistical tests in practice**

We have so far discussed the basic ideas of hypothesis testing in the context of one continuous and one discrete distribution. Before considering further tests, we shall look at some general matters concerning the application of hypothesis testing in practice.

The relationship between confidence intervals and tests
The reader may have noticed that the calculations required in the construction of a confidence interval for the mean, described in § 4.1, and in the setting up of a statistical test, described in § 5.1, are very similar. This is no coincidence, but a reflection of a close relationship between the confidence intervals and tests. In brief, if we can find a general method for constructing a $100(1 - \alpha)\%$ confidence interval for a parameter θ, say, then we can deduce a level-α test of the hypotheses $H_0 : \theta = \theta_0$ and $H_1 : \theta \neq \theta_0$ (where θ_0 is some specified value), and vice versa.

In more detail, suppose that (R, S) is a $100(1 - \alpha)\%$ confidence interval for θ, based on some data. Then a level-α test of $H_0 : \theta = \theta_0$ against $H_1 : \theta \neq \theta_0$ is

$$\text{'Reject } H_0 \text{ if } \theta_0 < R \text{ or } \theta_0 > S'. \tag{5.5.1}$$

For example, suppose that we find that a 95% confidence interval for the mean μ of a population is $(50.6, 52.3)$ and that we wish to test at the 5% level the hypothesis $H_0 : \mu = 50$ against $H_1 : \mu \neq 50$. We can conclude immediately that H_0 can be rejected at the 5% level since the value 50 lies outside the 95% confidence interval.

Exercise
Recall from § 4.1 that if $X_1, X_2, ..., X_n$ is a random sample from the $N(\mu, \sigma^2)$ distribution, with known σ^2, then a 95% confidence interval for μ is

$$(\bar{X} - 1.96\sigma/n^{1/2}, \bar{X} + 1.96\sigma/n^{1/2})$$

(1.96 being the 97.5% point of the $N(0, 1)$ distribution). Result (5.5.1) says that the test

$$\text{Reject } H_0 \text{ if } \mu_0 < \bar{X} - 1.96\sigma/n^{1/2} \text{ or } \mu_0 > \bar{X} + 1.96\sigma/n^{1/2}$$

of $H_0 : \mu = \mu_0$ against $H_1 : \mu \neq \mu_0$ has significance level α. Check that this agrees with the critical region given in table 5.1.

Exercise

The following voltages were recorded for a random sample of 6 batteries:

$$6.3, \quad 6.2, \quad 6.3, \quad 5.9, \quad 6.1, \quad 5.8.$$

Assuming that the voltages follow a normal distribution with variance equal to 0.04, show that a 90% confidence interval for μ, the mean voltage, is (5.97, 6.22). Test at the 10% level the hypothesis $H_0 : \mu = 6.0$ against $H_1 : \mu \neq 6.0$.

Proving that H_0 is true

We noted in § 5.1 that failure to reject the Null Hypothesis cannot by itself be used as evidence that the Null Hypothesis is true. Failure to reject might be a consequence of the Null Hypothesis being true or it might be that the Null Hypothesis is false but insufficient data have been collected to accumulate significant evidence against it. You may feel that this is rather unfair on the Null Hypothesis: although there is a chance of there being significant evidence against it, there is no possibility of significant evidence for it. The reason for this is that since the Null Hypothesis is of the form $\theta = \theta_0$, the whole population would need to be sampled to confirm its truth.

This difficulty can be eased if we can specify a range of values around θ that constitute differences of no practical importance. This range is called the 'specification range' in some contexts. For instance, in the tyre example of Illustration 5.1, the manufacturer may decide that if the mean life length of the new type of tyre lies in the range 39 to 41, then 'to all intents and purposes' they are equivalent to the old tyres, for which the mean was 40. Then one way of establishing that there is strong evidence that the Null Hypothesis is 'effectively' correct is to show that a confidence interval for the mean lies within the specification range. So, if the 95% confidence interval were (39.2, 40.6), then the manufacturer could conclude with confidence that the new process was 'practically equivalent' to the old—at least in terms of mean life length. On the other hand, if the confidence interval were (40.8, 43.2), say, then, by the relationship between confidence intervals and tests discussed above, the manufacturer could conclude that there was significant evidence at the 5% level against the Null Hypothesis (assuming $H_1 : \mu \neq 40$). An interval such as (39.3, 41.8) would be inconclusive.

One- and two-sided tests

The choice of the form of the Alternative Hypothesis is not always obvious. Suppose, for instance, that a pharmaceutical company has produced a new drug which it hopes will be more effective for curing a disease than the standard treatment. The proportion of sufferers of the disease recovering when the standard treatment is applied is known to be 0.6. Let p be the proportion recovering for the new drug. On the basis of the results of a clinical trial, it is required to draw inferences about the value of p. Should the Alternative Hypothesis be $H_1 : p > 0.6$ or $H_1 : p \neq 0.6$?

The answer is debatable, but the usual practice is to use the two-sided alternative, even though only values of p greater than 0.6 are sought. The reason for this choice appears to be that since the results of such a trial will be of general interest, other laboratories are likely to want to check them out. If the Alternative Hypothesis $H_1 : p > 0.6$ were used regularly, results apparently unfavourable to the new drug would be 'not significant' and their chance of being reported reduced. If $H_1 : p \neq 0.6$ is used, results both favourable and unfavourable to the new drug may prove 'significant', the experimenters reporting, of course, in which direction significance was achieved.

In short, one-sided tests may be appropriate for making purely 'internal' decisions, while two-sided tests are often preferred for results requiring 'outside' verification.

Costs, hunches and deferred decisions

One obvious missing factor in the tyre example of Illustration 5.1 is any consideration of costs or potential profits. Clearly in practice these are going to play a large role in deciding the number of new tyres to be made and tested, and, thereafter, whether the new method is to be adopted. The theory that we have developed can be modified to take account of the economic consequences of actions and an 'optimal' decision can be computed. The analysis of this type of problem comes under the heading 'decision theory'.

Another factor that we may want to incorporate is prior evidence or beliefs about the value of the parameter of interest. In the tyre example, for instance, if the manufacturer has a strong initial feeling that the new process is better than the old (perhaps based on theoretical or small-scale studies), he is more likely to be prepared to introduce it on the basis of 'marginal' test results than if he had no strong feelings initially. The amalgamation of 'hard' experimental results and prior or 'subjective' beliefs to produce rational decisions about the value of a parameter is achieved by making use of 'Bayesian statistical inference'.

Yet another natural modification to Illustration 5.1 would have been to allow the additional conclusion 'results inconclusive, test more tyres'. A possible complete scheme might then have been as follows: 'test 10 tyres; if the sample mean life length is less than 41, retain the old method, and if the sample mean exceeds 43, introduce the new method; if the sample mean lies between 41 and 43, test another 10 tyres and only introduce the new method if the sample mean for all 20 tyres exceeds 42'. In this way, we could reduce the chance of making a wrong decision while not increasing unduly the average number of tyres to be tested. Such a procedure is called a 'sequential test'.

The topics of decision theory, Bayesian statistical inference and sequential tests are important and interesting, but take us beyond the scope of this text. (An introduction to these topics may be found, for example, in *Introduction to the Theory of Statistics*, by A M Mood, F A Graybill and D C Boes (see the Bibliography)).

Choice of significance level
The choice of an appropriate significance level may cause difficulty. You can avoid having to make this choice by evaluating the exact significance, but if you are using statistical tables this may be difficult, especially for distributions other than the normal. One approach that is regularly adopted in this case is to test at the 5% level, then, if there is significance at this level, to test at the 1% level and then, if there is significance at the 1% level, to test at the 0.1% level. The significance is summarised as one of '$P > 0.05$', '$P < 0.05$', '$P < 0.01$' or '$P < 0.001$' depending on whether none, one, two or all three tests indicate significance.

⟩ Chapter 6

⟩ Statistical Tests: Applications

In Chapter 5 the basic principles of hypothesis testing were introduced in two very simple contexts: a normal distribution with known variance, and a binomial distribution. In this chapter we shall apply these principles to some more common situations. In § 6.1 we consider tests for one or two samples from normal distributions where the variance is unknown. In § 6.2 we go one stage further and see how so-called 'non-parametric' tests can be constructed; these tests require only very weak assumptions about the distribution of the data. Section 6.3 deals with the analysis of paired data, where there is an approximate linear relationship between the members of the pairs. Sections 6.4 and 6.5 both deal with types of χ^2 test: in the first case for the analysis of frequency data presented in a two-way table, and in the second to test whether a hypothesised family of distributions can provide a reasonable description of a set of data.

⟩6.1 One-sample, two-sample and paired *t*-tests

The one-sample t-test
We shall first consider a test on the mean of a normal distribution, based on a single random sample of observations $X_1, \ldots, X_n$. In Chapter 5, a test was derived under the assumption that the variance was known. Here we will concentrate on the more usual case of unknown variance.

Illustration 6.1
At one stage in an industrial chemical process, raw materials are fed into a vat where a reaction is initiated and, on average, 2 tons of output are

produced. After replacement of some of the machinery, the system has been reset and the following output weights were recorded for 12 reactions. The engineers wish to know whether the average output remains at 2 tons.

Weights (in tons) of output from a chemical reaction

2.011	2.018	1.986	2.032
2.003	2.025	2.019	2.013
1.995	2.011	2.006	2.002
1.992	2.014	2.002	2.009

Use the EDITOR to enter these data into a file, and then run program TESTS1 on the data. (You can also enter program TESTS1 with the default data, ignoring the initial picture and listing drawn by the program, and enter -999 in the editing facility. This will clear the default data and new data may then be entered.) A line diagram is drawn, from which we can see that the point 2.000 is certainly within the scatter of the data. We will proceed to test whether the mean output weight is 2.000 or not, assuming the data to be normally distributed but without assuming the variance to be known.

Select the 't-test' option; you will be prompted to enter the 'Value of μ under H_0'. The Null Hypothesis is that the mean is 2, so enter '2'. An arrow will mark the position of 2 on the line diagram. The hypotheses of interest, and future calculations, will be displayed in a coloured box on the screen. Our hypotheses are

$$H_0 : \mu = 2 \qquad \text{against} \qquad H_1 : \mu \neq 2.$$

Notice that the Alternative is assumed to be of the $\neq$ form. This is appropriate for the present example because if the weight has shifted from 2 we do not know whether it will be higher or lower. Now we have to consider how a suitable test might be constructed. We saw in Chapter 5 that, when the variance is known, a test can be based on the value of $\bar{X}$. This is because the distribution of $\bar{X}$ under H_0 is known; it is $N(\mu_0, \sigma^2/n)$, where μ_0 is the value of μ under the Null Hypothesis. In the present context, however, where σ^2 is not known, the distribution of $\bar{X}$ is not fully known either, so that a critical region cannot be found in the same way as before. Fortunately, we can make use of the sample variance S^2 and, instead of X, use the statistic

$$T = \frac{\bar{X} - \mu_0}{S/n^{1/2}}.$$

Result (4.2.3) tells us that if μ_0 is the true mean of the X_i, then T has a t_{n-1} distribution. This means that when H_0 is true,

$$T \sim t_{n-1}.$$

When the Null Hypothesis is true, $\bar{X}$ should have a value near μ_0 and so the statistic T will have a value near 0. When the Null Hypothesis is false, $\bar{X}$ should have a value further from μ_0 and so T will tend to be further from 0. In this way, T quantifies any evidence against the Null Hypothesis. In this context $T = (\bar{X} - \mu_0)n^{1/2}/S$ is called the *test statistic*.

Now it is time to calculate the observed value of our test statistic, which we can easily do by first noting the values of the usual summary statistics, $\bar{X}$, S and the sample size n. These will be calculated for you by the computer if the space bar is pressed twice. Another two presses produce the observed value of the test statistic, 2.869, and the significance level associated with this. Using t-tables (or with the aid of program DISTN), we find that the probability of a t_{15} random variable taking a value greater than 2.869, or less than -2.869, is 0.012. We therefore have quite strong evidence that the mean weight is no longer 2 tons, since the probability of obtaining a test statistic as extreme as the observed one is very small. Since the observed sample mean is larger than 2, we have quite strong evidence that the mean has shifted to a value larger than 2.

Worked example: one-sided test
The test described above is referred to as *two sided* because we have to examine the test statistic for departures from zero in either direction, i.e. we look for large positive values *or* large negative values. However, if the Alternative Hypothesis were $\mu < 2$, we would be interested only in large negative values. Similarly, if the Alternative Hypothesis were $\mu > 2$, we would be interested only in large positive values. In both these cases we would refer to the test as being *one sided*. Program TESTS1 always assumes that a two-sided test is to be carried out, but it is easy to modify the output of the program in order to carry out a one-sided test, as the following example shows.

Suppose that the tyre manufacturer introduced in Illustration 5.1 has collected a sample of tyre lifetimes and that he wishes to test the Null

Hypothesis that the mean is 40, with the Alternative that the mean has increased. This time, however, he does not wish to assume that the variance is known. The data are as follows.

41.7   41.1   40.2   41.5   42.4   39.2   39.9   41.2   40.5   40.7.

Use the EDITOR to create a file containing these data, or use the 'Edit the data' option in program TESTS1 to enter the data directly. (Remember that the case number -999 will clear all the present data.) Now select the '*t*-test' option and follow through the calculations. With the exception of the Alternative Hypothesis, all the other statements are correct until we arrive at the last line, which gives the significance of the observed test statistic. The test statistic still has a t_{n-1} distribution. However, it is now only large positive values of T that suggest that the Null Hypothesis should be rejected in favour of the Alternative. The observed value of T is 2.831. The printed significance (0.020) is therefore the probability that T lies above 2.831 or below -2.831 when H_0 is true. However, we wish to calculate only the probability that T lies above 2.831. Since the t distribution is symmetric, the probability of T lying below -2.831 is exactly the same as the probability of T lying above 2.831. The significance level for the one-sided test is therefore one half of the printed value, namely 0.01.

What would the significance have been if the observed value of T had been -2.831? Remember that we are interested in the probability lying above the observed value. We know that probability 0.020/2 lies below and so there is probability $1 - 0.020/2 = 0.99$ lying above. The significance level is therefore 0.99.

We have assumed throughout that the underlying distribution is normal; this allowed us to make an exact statement about the distribution of our test statistic. However, even if the underlying distribution is not normal, the *t*-test may still be appropriate. This is because the central limit theorem tells us that the sample mean $\bar{X}$ will be approximately normally distributed if the sample size is not too small. Consult §§ 2.7 and 4.3 for some discussion of this point.

The two-sample t-test
So far we have considered data only in the form of a *single* sample of observations. However, experimental data will often be more complicated than this, and may consist of several samples. In this section we will consider the case of two samples; the beetle data of table 1.2 on page

12 are an example of this kind. A natural question to ask is whether there is any evidence of a difference in the means of the two underlying populations. (We may have in mind the possibility that the length of a beetle might be used to determine its sex, if this is not easily identifiable by other means.) One possibility, then, is to derive a test to assess whether the male and female beetles have the same average body lengths.

Similar problems arise in many other contexts. As a second example, suppose that we are interested in comparing the effects of a new drug for a certain medical condition with the standard treatment. If ethical considerations allow it, then we might allocate a group of randomly chosen patients to a course of therapy on the drug and a second randomly chosen group to the standard treatment. After a suitable length of time, the strength of the disease might be measured in each patient. This would yield two samples of data, one from each group of patients. The question of interest is whether the strength of the disease in patients who are on the course of drugs is lower on average than the disease strength in patients on the standard treatment.

In general, the data to be analysed will be denoted by

$$X_1, ..., X_{n_x}$$
$$Y_1, ..., Y_{n_y}$$

where the use of X and Y distinguishes the two groups, and, in particular, n_x and n_y denote the numbers of observations in each sample. We will assume that each sample of data arises from a normal distribution, and we will also assume that the variances of these two distributions are the same. The case of different variances will be discussed later.

Run program TESTS2, which has the beetle data as default. There are two graphical displays available on this program, line diagrams and boxplots. You may like to use these to examine the data informally. (You may recall that the beetle data were used as an illustrative example for both of these techniques in Chapter 1.) In particular, there is no strong indication that the variances of the two populations are markedly different. Now select the 'Two sample t-test' option. Here, our hypotheses are

$$H_0 : \mu_x = \mu_y \qquad \text{against} \qquad H_1 : \mu_x \neq \mu_y$$

where μ_x and μ_y denote the population means of the two groups. Our first task is to construct a suitable test statistic. Let $\bar{X}, S_x, n_x$ denote the sample mean, standard deviation and sample size for males (X) and $\bar{Y}, S_y, n_y$ the same summary statistics for females (Y). We are interested in the difference between the population means μ_x and μ_y. A sensible starting point would therefore be to consider the difference in sample means $(\bar{X} - \bar{Y})$. From result (2.7.8), we know that $\bar{X} \sim N(\mu_x, \sigma^2/n_x)$ and $\bar{Y} \sim N(\mu_y, \sigma^2/n_y)$, where σ^2 denotes the variance of each population. Furthermore, we can appeal to a result similar to (2.7.9) to show that $\bar{X} - \bar{Y} \sim N(\mu_x - \mu_y, \sigma^2/n_x + \sigma^2/n_y)$. In order to estimate σ^2 we shall use the *pooled sample variance*, which is defined as

$$S_p^2 = \frac{(n_x - 1)S_x^2 + (n_y - 1)S_y^2}{n_x + n_y - 2}.$$

This combines the sample variance of each group into a single estimate of σ^2. If we now follow through the argument that we used in the case of the one-sample test, then we are led to consider the test statistic

$$T = (\bar{X} - \bar{Y})\left[S_p^2\left(\frac{1}{n_x} + \frac{1}{n_y}\right)\right]^{-1/2}.$$

Again, T will take values near 0 when the Null Hypothesis is true and values further from zero when it is false. Specifically, it can be shown that, when the Null Hypothesis is true,

$$T \sim t_{n_x + n_y - 2}.$$

Having selected the 't- test' option in program TESTS2, press the space bar to produce the values of the summary statistics for each group. You should use paper and pencil to follow through the calculations of the program as the results are produced on the screen. The observed value of the test statistic can be calculated quite easily from the displayed summary statistics. It has the value 2.221. Now, finally, the significance of the test statistic can be found by calculating the probability that an observation from a t_{21} distribution is greater than 2.221 or less than -2.221. With the beetle data, we have a significance level of 0.038. This provides quite strong evidence that males and females do differ in body length, with the males on average being slightly larger.

The test is derived under the assumption that the underlying distributions are normal. As in the one-sample t-test, the critical property is that the sample means $\bar{X}$ and $\bar{Y}$ are normally distributed, and we can sometimes appeal to the central limit theorem to use the test when the underlying distributions are not normal.

Exercise
Carry out a two-sample t-test on the following data, which refer to the

blood pressures of two groups of patients suffering from a certain medical condition. One group was given a drug designed to reduce blood pressure. The other was given a 'placebo', a treatment with no medical action but which allows the effect of administering a treatment to a patient to be monitored. (If we do not have such a 'control' group, we cannot say whether any observed change in blood pressures of those patients on the drug is due to the drug or to a psychological reaction of the patient to the fact that he is being treated.)

Diastolic blood pressures of patients (in mm of Hg)					
Control			Drug		
95.2	92.2	89.3	91.6	94.5	98.3
97.4	94.6	89.8	93.2	96.4	89.2
93.5	98.4	96.2	93.5	89.7	91.5
91.7			88.1		

Carry out this test on your own and then check your answer by using program TESTS2.

Notice that the tests in programs TESTS1 and TESTS2 always use Alternative Hypotheses of the $\neq$ form. On many occasions, a one-sided Alternative is more appropriate. Such a test can be performed by using these programs, with the exception of the final step where the observed significance level is calculated. If the displayed significance is p, the significance associated with the one-sided test is $p/2$ if the test statistic differs from zero in the direction of the Alternative Hypothesis or $1 - p/2$ if the test statistic differs from zero in the opposite direction. See the end of the previous section for a discussion of this.

The test we have derived assumes that the variances of the two populations are equal. If an informal examination of the data suggests that this assumption is unreasonable (standard textbooks describe a hypothesis test of equality of variances) then we can construct a slightly different test statistic, namely

$$T' = \frac{\bar{X} - \bar{Y}}{(S_x^2/n_x + S_y^2/n_y)^{1/2}}.$$

The distribution of T' cannot be found exactly but there are a number of good approximations. A simple one is that T' has approximately a t

distribution with degrees of freedom given by

$$df = \frac{(S_x^2/n_x + S_y^2/n_y)^2}{(S_x^2/n_x)^2/(n_x - 1) + (S_y^2/n_y)^2/(n_y - 1)}.$$

In order to use the t-tables, this number can be rounded to the nearest whole number. (This approximation is suggested in an article by B L Welch (1949 *Biometrika* **36** 293–6).)

The paired-sample t-test

A closer look at the previous exercise reveals some unsatisfactory features of the way the experiment was designed. We know that different patients will have different blood pressures because of unavoidable natural variation. Unfortunately, this variation will tend to obscure any differences between the placebo and drug groups. This happens because each observation has *two* sources of variation. First, each patient will have a different initial blood pressure because of the natural variation among patients. Secondly, we would expect that the application of each treatment will have slightly different effects on each patient. We are interested in the second source of variation. The first is simply a nuisance in that it will tend to blur any evidence of differences between the two groups.

A simple way of overcoming the first source of variation is to make a comparison of the effect of the two treatments on *each* patient. Suppose then that we design the experiment to examine 10 patients, and on each patient to record the blood pressure after the placebo is administered and also after the drug is administered. Notice that, in order to minimise any 'carry-over' effects, we should ensure that the order of administering the two treatments is chosen randomly in each case, and also that a sufficiently long time is allowed to elapse between the applications of the two treatments, so that any effect from the first has disappeared. Suppose that we carry out the experiment in this way and obtain the results listed in the table on page 136.

We now have two sets of measurements:

$$X_1, ..., X_n$$

$$Y_1, ..., Y_n.$$

However, each observation from one group is now *paired* with a corresponding observation from the other group through being taken on the same patient. This is an example of paired data.

Patient	Placebo	Drug
1	98.1	96.2
2	92.8	91.3
3	85.4	82.9
4	94.9	94.5
5	91.2	91.5
6	86.5	84.4
7	87.7	86.5
8	92.0	92.4
9	90.4	89.3
10	89.3	87.5

Run program TESTS1 and use the default data, which consist of the set of paired blood pressures just described. The data are represented in a line diagram as before. However, the pairing can now be represented by joining the paired measurements with straight lines. Notice that if the drug measurement is generally lower on each patient than the placebo measurement then the lines will tend to slope in the same direction. Different patients may have widely different blood pressures on the placebo, but each of them might drop by roughly the same amount when the drug is applied. So, by paying attention to the *difference* of the two measurements on each patient we will eliminate the variation due to the initial blood pressures of each patient and concentrate attention on the relative effects of the two treatments. This method of designing the experiment allows the effectiveness of the drug to be assessed in a more powerful way than by a two-sample t-test.

In order to proceed with a formal test we first replace the two sets of observations by the differences between the pairs. This reduces the data to the form

$$D_1, ..., D_n$$

where $D_i = X_i - Y_i$. With the present set of data, this creates the observations

$$1.9, \quad 1.5, \quad 2.5, \quad 0.4, \quad -0.3, \quad 2.1, \quad 1.2, \quad -0.4, \quad 1.1, \quad 1.8.$$

If there is no difference between the two treatments then these numbers should be a random sample from a distribution with mean zero. Specifically, our Null Hypothesis for a paired-sample t-test is $H_0 : \mu_d = 0$, where μ_d denotes the mean of the differences. We can therefore carry out tests on paired data by carrying out a one-sample test on the differences of the pairs.

Carry out a t-test on this set of data, using the hypotheses

$$H_0 : \mu_d = 0 \qquad \text{against} \qquad H_1 : \mu_d > 0$$

by selecting the 't-test' option in TESTS1. At the last step, remember that the significance level on the screen refers to the $\neq$ form of Alternative. Since the Alternative Hypothesis is now $\mu_d > 0$, the usual adjustment should be made to the printed significance in order to take account of this. Since the test statistic is greater than zero, the significance of the observed value of the test statistic is $0.005/2 \simeq 0.003$.

The paired-sample t-test is derived under the assumption that the differenced data come from a normal distribution. As usual, the test can be applied in an approximate way to data whose differences are not normally distributed if the sample size is large enough for the central limit theorem to take effect.

Table 6.1   Mathematical details of one-sample, two-sample and paired t-tests. These tests assume that the sample means have a normal distribution.

Type of data	Null Hypothesis	Test statistic	Alternative	Critical region for sig level α
One sample $X_1, ..., X_n$	$H_0 : \mu = \mu_0$	$T = \dfrac{\bar{X} - \mu_0}{S/n^{1/2}}$	$H_1 : \mu \neq \mu_0$	$T < -t_{n-1,1-\alpha/2}$ or $T > t_{n-1,1-\alpha/2}$
			$H_1 : \mu > \mu_0$	$T > t_{n-1,1-\alpha}$
			$H_1 : \mu < \mu_0$	$T < -t_{n-1,1-\alpha}$
Two samples $X_1, ..., X_{n_x}$ $Y_1, ..., Y_{n_y}$	$H_0 : \mu_1 = \mu_2$	$T = \dfrac{\bar{X} - \bar{Y}}{[S_p^2(1/n_x + 1/n_y)]^{1/2}}$	$H_1 : \mu_x \neq \mu_y$	$T < -t_{n_x+n_y-2,1-\alpha/2}$ or $T > t_{n_x+n_y-2,1-\alpha/2}$
			$H_1 : \mu_x > \mu_y$	$T > t_{n_x+n_y-2,1-\alpha}$
			$H_1 : \mu_x < \mu_y$	$T < -t_{n_x+n_y-2,1-\alpha}$
Paired samples $X_1, ..., X_n$ $Y_1, ..., Y_n$	$H_0 : \mu_d = 0$	$T = \dfrac{\bar{D}}{S_d/n^{1/2}}$	$H_1 : \mu_d \neq 0$	$T < -t_{n-1,1-\alpha/2}$ or $T > t_{n-1,1-\alpha/2}$
			$H_1 : \mu_d > 0$	$T > t_{n-1,1-\alpha}$
			$H_1 : \mu_d < 0$	$T < -t_{n-1,1-\alpha}$

⟩6.2   Non-parametric tests

There are occasions when we might be unhappy to assume that our data are normally distributed. We may even be reluctant to assume that the sample means are approximately normal, perhaps because the sample size is too small for the central limit theorem to take effect. In such situations, we might prefer to use a test which is not based on a normal assumption. Three such tests, covering one-sample, two-sample and paired-sample data, are derived in this section.

There are also occasions when we are forced to abandon any assumption of a particular continuous probability distribution for the data. This happens when the data consist of a set of *ranks*, so that we know only the ordering of the observations with respect to one another. The data cannot then be interpreted as measurements on a numerical scale in the usual way. For such data, we have to use the ranks to derive the tests we need.

The sign test

This test is used when we have a single sample of data and wish to test a hypothesis about the centre of its distribution. In order to demonstrate the sign test we shall re-analyse the chemical yield data of Illustration 6.1, but this time without making the assumption that the sample mean has an approximate normal distribution. We must first consider carefully the nature of our hypotheses. In the t-test, the Null Hypothesis was that the mean took the value 2. However, if we are to abandon the assumption of normality and use only the ranks of the data then it is difficult to deal with means since we no longer have the data to evaluate them or the scale of measurement to define them. Fortunately, we can still refer to medians because they are defined in terms of ranks. For example, the median of a group of nine observations is the value of the fifth observation. For this reason, all the tests derived in the present section will be concerned with hypotheses on medians. The population *median* of a distribution, namely the 50% point, will be denoted by m.

Run program TESTS1 on the chemical yield data, select the 'Sign test' option and enter '2' for the value of the median under the Null Hypothesis. Our hypotheses are therefore

$$H_0 : m = 2 \qquad \text{against} \qquad H_1 : m \neq 2.$$

The first operation that is carried out on the data is an ordering of the values. If 2 is the median then we would expect approximately half the data to lie below 2 and approximately half to lie above. This suggests that a suitable test statistic would be

T = number of observations which are greater than 2.

If the Null Hypothesis is true, T should be near $n/2$. Very large or very small values of T provide evidence against the Null Hypothesis.

In order to evaluate the significance of the observed test statistic, we need to find the distribution of T under the Null Hypothesis. We have n independent observations in our sample and, by definition of the median of a distribution, the probability of each one taking a value larger than 2 is $\frac{1}{2}$. So, under the Null Hypothesis, the number of observations larger than 2 has a binomial distribution with parameters n and $\frac{1}{2}$, i.e.

$$T \sim B(n, 0.5).$$

Follow through the calculations by pressing the space bar repeatedly and note that the observed value of T is 13. Make sure that you understand what is happening at each step as it is illustrated. Unfortunately, there is only enough space on the screen to display the test in summary form. You should write out the full details for yourself on a piece of paper. At the last step the significance of the test statistic is displayed as 0.024. In order to see how this is calculated, run program BINTEST, setting the binomial probability to 0.5 and the sample size to 16, and choosing the Alternative Hypothesis of the '$\neq$' form. When the 'Evaluate significance' option is selected, the distribution of T under the Null Hypothesis is displayed. Enter 13 as the observed value. The significance is the probability of obtaining a value as, or more, extreme than 13 when H_0 is true. As the space bar is pressed, the region corresponding to values of 13 and higher is shaded and the probability associated with this region is evaluated. However, we also need to consider the probability associated with the corresponding extreme values from 3 down to 0 since it is both very large or very small values that will lead us to reject H_0. Since the binomial distribution is symmetric when the parameter $p = \frac{1}{2}$, this second probability is in fact equal to the first one. The significance of our observed test statistic is therefore $2 \times 0.0105 = 0.021$. This calculation can be checked by selecting the 'Binomial probabilities' option.

The value 0.021 was derived by an exact calculation. In fact, program TESTS1 uses a normal approximation to binomial probabilities, with a continuity correction, as discussed in § 2.7. This can be illustrated in program B&P by selecting the 'Normal approximation' option, listing the probabilities and summing the appropriate values once more. In this case the two agree quite closely since the normal approximation produces an approximate significance of 0.024. In general, the normal approximation used in the sign test is

$$T \doteqdot N(n/2, n/4).$$

All the non-parametric tests to be discussed in this section are implemented on the computer with normal approximations. This means that care should be taken in using the programs when the sample sizes are very small, although the normal approximations are in general very good. Critical regions for each of the non-parametric tests, for sample sizes up to 15, are listed in tables 5, 6 and 7 of Appendix 4. In the case of the sign test, the exact calculations can be carried out with the aid of program BINTEST, but, for convenience, critical regions are tabulated in table 5. This means that we can carry out a sign test at some prespecified significance level, simply by finding the observed value of the test statistic and then referring to table 5.

The sign test is an example of a 'non-parametric' test because it dispenses with the assumption that the distribution of the data has some known shape whose exact form is governed by a small number of unknown parameters. The test succeeds in doing this by ignoring some information in the data, namely the exact location of the data points. The only information that is used is whether each data point is smaller or larger than the value of the median proposed under the Null Hypothesis. This means that the test will not have as much power as the t-test when the data really are normally distributed. (We are using the term 'power' in the technical sense, described in § 5.3.) This can be seen to a small extent by the fact that a sign test on the chemical yield data produces a significance of 0.021, whereas the t-test on the same data produced a significance of 0.012. The result of the t-test is slightly more conclusive than the sign test. So, if we correctly assume that the data are normally distributed then we will have a slightly more powerful test. However, if we assume that a normal distribution is appropriate when in fact it is not then our test may be slightly inaccurate, and at worst positively misleading. This is the choice that we have to make between the two tests.

Worked example: observations tied with the median
Suppose that the observation 2.002 in the chemical data had actually taken the value 2.000. Use the 'Edit the data' option in the program to

change this value and then rerun the sign test. Now we have an observation which happens to take the exact value of the proposed median. The simplest way of dealing with such a problem is to remove the observation from the data for the purpose of carrying out the test, since it is neither greater than or less than the proposed median. Follow through the details of the test on the screen and notice that the value of n is reduced by 1.

Exercise: one-sided tests
Consider a sign test applied to the tyre manufacturer's data in § 6.1. Program TESTS1 calculates the significance under the assumption of a two-sided Alternative. How would you evaluate the significance in the one-sided case? Use program BINTEST to assist your thoughts about this.

Exercise: paired data
The sign test is often used in the context of paired data. Work through a sign test of the blood pressure data, which are the default data in program TESTS1. Assume that the Alternative Hypothesis is one sided, i.e. interest lies in whether the diastolic blood pressure has been reduced by the drug.

The Wilcoxon signed ranks test
There is a test which can improve on the power of the sign test if we are prepared to assume that the underlying distribution is symmetric about its median. Run program TESTS1 with the chemical data again, select the 'Signed ranks test' option and enter '2' as the value of the median under the Null Hypothesis. We are considering again a test of the hypotheses

$$H_0 : m = 2 \qquad \text{against} \qquad H_1 : m \neq 2.$$

Press the space bar twice to see the proposed median subtracted from each observation and the data sorted. However, this time the sorting is done with respect to the absolute value of the difference of each observation from 2. In other words, the differences are ordered from smallest to largest, ignoring the signs. The correct signs are, however, retained on each number. The idea behind this manipulation is to see where the negative differences lie with respect to the positive ones. If we look at the *number* of positive values, then we are essentially carrying out a sign test on the data. However, we can also make use of the additional information contained in the *ranks* of the positive numbers among the entire set. The ranks are the coloured (or shaded) numbers in the left-hand column, which replace each observation by an integer referring to its order in the list. If the positive numbers tend to have higher ranks then this will reinforce the evidence against the Null Hypothesis.

A problem sometimes arises in that two or more observations take the same value and so we have *ties*. With the present data, the first and second observations are tied, as are the eighth and ninth and the eleventh and twelfth. This is resolved simply by assigning to each of the tied observations the average of the corresponding set of ranks.

A suitable test statistic is based on the ranks of the observations which have positive signs. Specifically, the signed ranks test has test statistic

$$T = \text{sum of the ranks of the positive differences.}$$

Press the space bar to see the signs of the observations passed to the ranks. If the Null Hypothesis is true then we would expect the negative ranks to be distributed evenly, although randomly, throughout the entire list. If the Null Hypothesis is false then we would expect the positive ranks to outweigh the negative ones, or vice versa. In this way, we see that unusually large or unusually small values of the statistic will lead us to reject the Null Hypothesis. The distribution of T under the Null Hypothesis can be computed exactly; critical regions of the test for sample sizes up to 15 are listed in table 6 of Appendix 4. (For example, the rejection region for a two-sided test at the 5% level, with sample size 7, consists of the values $T \leqslant 2$ or $T \geqslant 26$.) Program TESTS1 uses the normal approximation

$$T \doteq N(n(n + 1)/4, n(n + 1)(2n + 1)/24).$$

Repeated pressing of the space bar causes the signs of the observations to be passed to the ranks, the ranks with negative signs deleted and the remaining ranks summed. A further press of the space bar shows the resulting test statistic to be slightly more significant than in the sign test, reflecting the fact that the signed ranks test is a slightly more powerful test.

This test is sometimes referred to as the Wilcoxon signed ranks test after the statistician who first proposed it. The assumption of symmetry of the underlying distribution is necessary since features such as skewness can produce 'significant' test statistics even when the true value of the median is the one indicated by the Null Hypothesis.

Exercise: paired data
Like the sign test, the signed ranks test is often applied in the context of paired data. Work through a signed ranks test of the paired blood

pressure data, which are the default data in program TESTS1. Again, remember that a one-sided test is the most natural one to carry out here.

The Mann–Whitney U-test

Consider the two-sample problem posed by the beetle data of table 1.2 and run program TESTS2 for illustration. In this case we have data of the form

$$X_1, ..., X_{n_x} \qquad Y_1, ..., Y_{n_y}.$$

The hypotheses of interest are

$$H_0 : m_x = m_y \qquad \text{against} \qquad H_1 : m_x \neq m_y$$

where m_x and m_y denote the medians of the X and Y populations respectively. We assume that the distributions of X and Y have the same shape and we wish to test whether the medians are identical.

Select the 'Mann–Whitney U-test' option and press the space bar repeatedly to see an illustration of a non-parametric test in this context. The data in each group are first sorted from smallest to largest. A column of ranks from 1 to n ($= n_x + n_y$, the total number of observations) is then printed in the centre of the screen. Next, each observation travels in turn across the screen to its allotted place in this central column of ranks. In this way, the entire collection of data is ranked from 1 to n but the group membership of each observation remains identifiable through the side of the column on which the number appears. If the Null Hypothesis is true and the medians of the two populations are identical then we would expect to see the observations from the two groups intermingled with one another with no marked patterns or clusters. Specifically, if observations from one group tend to gather at one end of the list and the observations from the other group at the other end then this would provide evidence against the Null Hypothesis.

A useful test statistic can therefore be based on the values of the ranks allocated to the two groups. Repeated pressing of the space bar produces a demonstration of this by separating the two sets of ranks and then summing them. As usual, if ties occur these can be resolved by assigning average ranks. For example, two observations are tied at ranks 16 and 17 and these are allocated the average rank 16.5. Similarly, the observations tied at positions 18 and 19 are each allocated the rank 18.5. A possible test statistic is

$$T_x = \text{the sum of the ranks of the } X\text{'s}$$
$$\text{in the ordered collection of all the data}$$

or we might use

$$T_y = \text{the sum of the ranks of the } Y\text{'s}$$
in the ordered collection of all the data.

However, it is more convenient to consider the equivalent statistics

$$U_x = n_x n_y + n_x(n_x + 1)/2 - T_x$$

and

$$U_y = n_x n_y + n_y(n_y + 1)/2 - T_y.$$

(It can be shown that these statistics record the number of times an X precedes a Y, and vice versa, in the ordered list.) Our final statistic combines these into

$$U = \text{smaller of } U_x \text{ and } U_y.$$

It is traditional to refer to this statistic as U rather than T and to refer to the test as the Mann–Whitney U-test, after the statisticians who were among the first to propose and study it.

Small values of U will lead us to reject the Null Hypothesis of equal population medians. With sufficient work, it is possible to derive the exact distribution of this statistic when the Null Hypothesis of equal medians is true; critical regions based on this exact distribution are given in table 7 of Appendix 4. Program TESTS2 employs the normal approximation

$$U \doteq N(n_x n_y/2, n_x n_y(n_x + n_y + 1)/12).$$

Complete the details of the test on the beetle data by pressing the space bar and check the details of your test by using the critical regions provided in table 7.

This test does not assume that the underlying distributions are normal. It does, however, assume that the shapes of the two underlying distributions are identical. Again, we should expect this test to be slightly less powerful than the two-sample t-test when the data are normally distributed.

Exercise
Carry out a Mann–Whitney U-test on the two samples of blood pressure data on page 134. (Do not use the second set of blood pressure data, for which a paired test is appropriate.) Use the Alternative Hypothesis $\mu_x > \mu_y$. Carry out the test with paper and pencil and check your answer by using program TESTS2. You will have to adjust the significance level

printed on the screen since a one-sided Alternative Hypothesis is being assumed. For a one-sided test, it is necessary to take into account which of U_x and U_y produced the smaller value when deciding whether there is sufficient evidence to reject the Null Hypothesis.

Table 6.2 Mathematical details of one-sample and two-sample non-parametric tests. Paired data may be analysed by carrying out a sign or signed ranks test on the differences $X_i - Y_i$. Some critical regions for these tests are listed in tables 5, 6 and 7 of Appendix 4.

Type of data	Name of test	Null Hypothesis (m refers to the median)	Test statistic	Distribution under H_0
One sample $X_1, ..., X_n$	Sign test	$H_0 : m = m_0$	T = number of observations $> m_0$	$B(n, \frac{1}{2})$ (exactly) $N(n/2, n/4)$ (approximately)
One sample $X_1, ..., X_n$	Wilcoxon signed ranks test	$H_0 : m = m_0$ and distribution is symmetric	T = sum of the ranks of the positive differences, $X_i - m_0$	$N\left(\dfrac{n(n+1)}{4}, \dfrac{n(n+1)(2n+1)}{24}\right)$ (approximately)
Two samples $X_1, ..., X_{n_x}$ $Y_1, ..., Y_{n_y}$	Mann–Whitney U-test	$H_0 : m_x = m_y$ and the two distributions have the same shape	Let T_x and T_y denote the sum of the ranks of the first and second groups respectively U = smaller of $n_x n_y + n_x(n_x + 1)/2 - T_x$ and $n_x n_y + n_y(n_y + 1)/2 - T_y$	$N\left(\dfrac{n_x n_y}{2}, \dfrac{n_x n_y (n_x + n_y + 1)}{12}\right)$ (approximately)

⟩6.3 Simple linear regression

The tests that we have discussed so far in this chapter have been concerned with problems involving one sample, two samples or paired samples of data. In this section we will again consider pairs of measurements but the main interest now lies in exploring the nature of the relationship between the variables which make up the pairs. We shall discuss this situation in the context of the following example, which was first discussed in § 1.1.

Illustration 6.2
A medical researcher is interested in whether (and, if so, how) the level
of a protein changes in expectant mothers throughout pregnancy. Observations have been taken on 19 women at different stages of pregnancy.

Time into pregnancy (weeks), x	Protein level (mg ml^{-1}), Y
11	0.38
12	0.58
13	0.51
15	0.38
17	0.58
18	0.67
19	0.84
21	0.56
22	0.78
25	0.86
27	0.65
28	0.74
29	0.83
30	0.99
31	0.84
33	1.04
34	0.92
35	1.18
36	0.92

Data of this type will in general be denoted by

$$(x_1, Y_1), (x_2, Y_2), \ldots, (x_n, Y_n).$$

The use of upper and lower case stresses the fact that the values of Y are
subject to random variation whereas the values of x are not.

Run program SCATTER with the default set of data, and a scatterplot
of the protein data will be drawn on the screen. The plot clearly indicates
an increase in the protein level over time. We might plausibly also argue
that the underlying relationship between the two variables is approximately linear, with a certain amount of 'noise' superimposed.
Mathematically, we can express this in a model as follows:

$$Y_i = \alpha + \beta x_i + \varepsilon_i.$$

This kind of model is referred to as a *regression* model. It describes the behaviour of one variable Y in terms of a linear function of another variable x. The expression $\alpha + \beta x$ describes a straight line, where α is the intercept (the height at which the line crosses the vertical axis $x = 0$) and β is the slope (or rate of increase). The model assumes that the error term is simply added on. It is also assumed that each error term has mean zero and has the same variance, which we shall denote by σ^2.

The object of this experiment is to see how one variable (protein level) is affected by another (time). In this example, protein level is referred to as the *response* variable (which in general will be denoted by Y). By contrast, time is referred to as the *explanatory* variable (which will in general be denoted by x) since we expect changes in protein level to be 'explained' by changes in time. We must beware of implying through use of the term 'explanatory' that this variable is the cause of the response variable. There may be other variables involved which we have not measured and which are the real cause or driving force of the relationship under investigation. An alternative description of the two variables would be to say that time is an *independent* variable and protein level is *dependent* (on time).

Select the 'Regression' option and a line will appear on the screen. This represents the linear part of the model described above. As the 'cursor keys' (i.e. the four keys with arrows printed on them) are pressed, the line will move about the screen and the formula at the top of the screen will record the current values of the slope and intercept parameters. We are now faced with the problem of how to fit this regression model to our observed data.

In order to fit the model we need some sort of criterion by which to assess how well a proposed model describes the observed data. If the key R is pressed, a number of vertical lines are drawn from the data points to the current regression line. The sum of the squared distances of these lines provides a quantitative measure of the 'agreement' between the current line and the data. Notice that the vertical distances, as opposed to the horizontal distances, are used. This reflects the fact that for a given time the observed protein level is a linear function of time plus error. If the true regression line were the one displayed on the screen then the vertical bars would represent the errors. For any x_i, the corresponding point on the regression line is given by $\alpha + \beta x_i$, and so the squared length of the line from here to the corresponding data point is $(Y_i - \alpha - \beta x_i)^2$. Our criterion, called the *sum-of-squares function,* can be expressed in mathematical terms as

$$\mathrm{SS}(\alpha,\beta) = \sum_i (Y_i - \alpha - \beta x_i)^2.$$

The sum of squares is a function of the parameters α and β. We can therefore fit our regression model by choosing the values of α and β that minimise $\mathrm{SS}(\alpha,\beta)$. In other words, we are choosing our parameter estimates in such a way that the regression line 'agrees' with the observed data as closely as possible. Try moving the line about in order to achieve as small a sum of squares as you can. Notice that the numerical value of the sum of squares is printed at the top of the screen.

It would clearly be better to have an 'automatic' way of fitting models such as these, without having to go through the process of minimising the sum-of-squares function numerically. It can be shown that the values of α and β that minimise $\mathrm{SS}(\alpha, \beta)$ are given by

$$\hat{\alpha} = \bar{Y} - \hat{\beta}\bar{x} \qquad \hat{\beta} = S(x, Y)/S(x, x)$$

where

$$\bar{x} = \frac{1}{n} \sum_{i=1}^{n} x_i \qquad \bar{Y} = \frac{1}{n} \sum_{i=1}^{n} Y_i$$

$$S(x, x) = \sum_{i=1}^{n} (x_i - \bar{x})^2 \qquad S(x, Y) = \sum_{i=1}^{n} (x_i - \bar{x})(Y_i - \bar{Y})$$

$$= \sum_{i=1}^{n} x_i^2 - n\bar{x}^2 \qquad = \sum_{i=1}^{n} x_i Y_i - n\bar{x}\bar{Y}.$$

These convenient formulae make the fitting of regression models very easy. Press the F key to draw the fitted model on the screen. The display that should now be on the screen is illustrated in figure 6.1. How well did you do by eye? The sum-of-squares function is now at the smallest achievable value. This is given the special name of the *residual sum of squares* or RSS for short. The individual vertical lines, whose sum of squares produces the RSS, are referred to as the *residuals*.

There is one further parameter in the model which we have still to estimate, namely the variance σ^2 of the error terms ε_i. It can be shown that an unbiased estimator of σ^2 is $\hat{\sigma}^2 = \mathrm{RSS}/(n - 2)$, where n is the sample size ($n = 19$ in the present example). Sometimes the notation s is used in place of $\hat{\sigma}$. The RSS can be calculated conveniently through the formula

$$\mathrm{RSS} = S(Y, Y) - S(x, Y)^2/S(x, x)$$

where
$$S(Y,Y) = \sum_i (Y_i - \bar{Y})^2 = \sum_i Y_i^2 - n\bar{Y}^2.$$

A model of this sort is very useful in that we now have a formula, namely
$$\hat{Y} = \hat{\alpha} + \hat{\beta}x$$

which summarises the relationship between Y and x. Notice that $\alpha + \beta x$ is the *average* protein level corresponding to the time x. The estimate of this average protein level is denoted by $\hat{Y}$ and corresponds to a point on the fitted regression line. $\hat{Y}$ is referred to as a *fitted value*. In the protein example, we obtain the formula $\hat{Y} = 0.202 + 0.023x$. However, a formula of this type gives no indication of the scatter of errors about the regression line and so we should complete our summary with an estimate of the standard deviation of the error terms; in this case, $s = 0.115$.

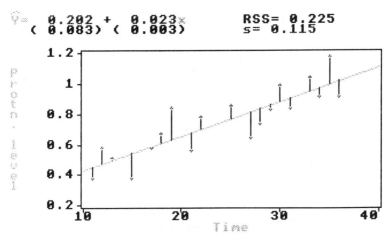

Figure 6.1 The least-squares regression line fitted to the protein data in program SCATTER, with the residuals displayed as vertical bars.

We should, as always, make some attempt to check that the fitted model describes the data well. For regression data we can do this simply by examining carefully the scatterplot of the data, preferably with the fitted line superimposed, in order to investigate possible non-linearities. Remember that we have also assumed the variance of the error terms to be constant, i.e. it is the same at all points on the regression line. However, it is sometimes easier to assess these assumptions by looking for trends in the residuals. A *residual plot* is constructed by plotting the residuals separately, against the corresponding values of the explanatory variable. Systematic trends in this plot would indicate possible departures from the proposed model. Press R to produce animation which illustrates the construction of such a plot.

It is instructive to consider how we would establish, in a more formal way than by examination of a scatterplot, that the average protein level does indeed change with time. This can be done by carrying out a test of the hypotheses

$$H_0 : \beta = 0 \qquad H_1 : \beta \neq 0.$$

In order to do this, we must first make some assumption about the distribution of the error terms ε_i. We will assume that this distribution is normal, so that $\varepsilon_i \sim N(0, \sigma^2)$. The assumption of normality can also be checked by examining the residuals, for example by constructing a histogram and comparing the shape of this with a normal density curve.

In order to carry out tests on the parameter β, we can now make use of the fact that when the model is correct,

$$\hat{\beta} \sim N(\beta, \sigma^2/S(x, x)).$$

Furthermore,

$$\frac{\hat{\beta} - \beta}{(\hat{\sigma}^2/S(x, x))^{1/2}} \sim t_{n-2}$$

and so a test of the hypothesis $\beta = 0$ is achieved by comparing the test statistic

$$\frac{\hat{\beta}}{\hat{\sigma}/(S(x, x))^{1/2}}$$

with a t_{n-2} distribution. In the present case, we have $\hat{\beta} = 0.023$, $\hat{\sigma}/(S(x, x))^{1/2} = 0.003$ and $n = 19$, giving the observed value of the test statistic as $0.023/0.003 = 7.7$. (We should in practice use a larger number of significant figures for such a calculation but the display has not provided them in this particular case. The computer does, of course, use a large number of significant figures in its own internal calculations.) The 97.5% point of the t_{17} distribution is $t_{17,0.975} = 2.11$. We can therefore say that there is significant evidence that the slope parameter is not zero. In other words, there is significant evidence that the protein level changes over time.

We saw in § 5.5 that the outcome of a hypothesis test can be deduced from the examination of an appropriate confidence interval. In the present case, a confidence interval for β is provided through the formula

$$(\hat{\beta} - t_{n-2,1-\alpha/2}(\hat{\sigma}^2/S(x, x))^{1/2}, \ \hat{\beta} + t_{n-2,1-\alpha/2}(\hat{\sigma}^2/S(x, x))^{1/2}).$$

Notice that the *standard error* of β (an estimate of the standard deviation of $\hat{\beta}$) is printed in brackets at the top of the screen, underneath $\hat{\beta}$; in the present case, this is simply the expression $\hat{\sigma}/(S(x, x))^{1/2}$. Knowledge of the standard error allows a confidence interval to be calculated very easily.

Exercise
Calculate a 95% confidence interval for the slope parameter in the protein data and verify that zero is not contained within it.

Exercise: confidence interval for the intercept parameter α
We may sometimes also wish to carry out a test on the value of α. This can be done in a similar way to the test on β, through the results

$$\hat{\alpha} \sim N(\alpha, \sigma^2(1/n + \bar{x}^2/S(x, x)))$$

and

$$\frac{\hat{\alpha} - \alpha}{(\hat{\sigma}^2(1/n + \bar{x}^2/S(x, x)))^{1/2}} \sim t_{n-2}.$$

In the protein data, it is possible that the protein level starts at zero when pregnancy begins. This suggests the simpler model $Y_i = \beta x + \varepsilon_i$. Construct a confidence interval for α in order to assess whether this simple model is a reasonable one.

Exercise
The following data refer to the times taken for a certain chemical reaction to occur in the presence of different amounts of catalyst.

Catalyst concn: x(mg g^{-1})	5	6	7	8	9	10	11	12	13	14
Reaction time: Y(s)	9.4	8.3	7.9	8.6	7.4	6.9	7.0	6.5	4.6	4.7

Fit a simple linear regression model to these data, with reaction time as the response and catalyst concentration as the explanatory variable, and find a 95% confidence interval for the slope parameter. Carry out the calculations by hand and then check your answers by entering the data into program SCATTER.

If the EDITOR is used to create a datafile which is then read into program SCATTER, a prompt will occur ('Which is the response variable? Enter 1 or 2') to identify which of the two columns of figures refers to the response variable. For example, if a file is constructed for the example above with the catalyst concentrations in the first column and the corresponding reaction times in the second, then '2' should be given in reply to the prompt.

Exercise: outliers and non-normality
Just as outliers can occur in single samples of data, so one or two suspect observations can cause problems in regression data. Run program SCATTER with the default protein data and select the 'Edit the data' option. Change case number 18 to read 35 for Time and 1.78 for protein level and replot the data. The altered observation 18 now looks out of place. Its protein level is rather high compared with the others. This could easily have been overlooked if the two columns of figures had been scanned by eye, but it is quite obvious from the plot.

If we had observed the data in this form we would be well advised to go back to the source of the data and check on this observation. For the moment, we could assess the influence that this observation would have on our analysis by fitting the model on the data as they stand and then refitting the model with observation 18 omitted. Do this and write down the parameter estimates and their standard errors in each case. You should find that observation 18 has rather a large influence on the results; this illustrates once more the importance of plotting the data. Notice that the influence and unusual nature of observation 18 is also reflected in the residual plot, which has a rather large residual for this observation, and neighbouring residuals, which are mostly negative.

Exercise: transformations
Use the EDITOR to create a file containing the data in the table below and then read it into program SCATTER.

The scatterplot of these data does not look at all linear. However, our linear regression model may still be appropriate if we can find a transformation of x or Y, or both, which will make the assumption of linearity plausible. Select the 'Transform Y' option and enter the function $LN(Y)$. The resulting scatterplot is very different and a linear assumption now looks quite reasonable. This shows the power of a suitably chosen transformation.

Try some transformations on the data in their present form and look

x	Y
0.17	2.7
3.05	7.4
4.97	20.1
4.79	54.6
4.33	148.4
6.81	403.4
8.35	1 097.2
10.24	2 981.0
9.25	8 103.6
10.63	22 026.4
11.25	59 874.9
12.52	161 021.1

at the effects on the plot. In each case, work out the transformation that will return the plot to its original form.

When carrying out transformations we must remember that linearity is not the only assumption that our model makes. We also assumed that the error variance was constant. This means that the scale of the scatter of the observations about the true regression line should be the same for all values of x. Any transformation which is used on the data must therefore produce a scatterplot for which these two assumptions are simultaneously plausible.

Table 6.3 Mathematical details for simple linear regression.

Model	$Y_i = \alpha + \beta x_i + \varepsilon_i \qquad i = 1, ..., n$		
Parameter estimates	$\hat{\alpha} = \bar{Y} - \hat{\beta}\bar{x}$	$\hat{\beta} = S(x, Y)/S(x,x)$	$\hat{\sigma}^2 = \text{RSS}/(n-2)$
Distribution, assuming $\varepsilon_i \sim N(0, \sigma^2)$	$N(\alpha, (1/n + \bar{x}^2/S(x,x))\sigma^2)$	$N(\beta, \sigma^2/S(x,x))$	—
Test statistic for zero parameter value	$\dfrac{\hat{\alpha}}{[(1/n + \bar{x}^2/S(x,x))\sigma^2]^{1/2}}$	$\dfrac{\hat{\beta}}{(\hat{\sigma}^2/S(x,x))^{1/2}}$	—
Distribution of the test statistic under Null Hypothesis	t_{n-2}	t_{n-2}	—

⟩6.4   Contingency tables

A contingency table is a table of frequencies. For instance, in an investigation of eye colour and quality of vision, the summarised data might be presented in a two-way table where the rows indicate eye colours, the columns indicate various levels of vision and the frequencies are the numbers of people having the different combinations of eye colour and level of vision. A natural question is then whether the distributions of levels of vision are the same for people with each eye colour. In statistical jargon, we are asking whether there is an *association* between eye colour and level of vision.

The test we are going to develop makes use of a family of distributions that we have not seen so far, the χ^2 distributions. (χ is the Greek letter 'chi', the 'ch' being pronounced as a 'k'.) Like the t distributions, the χ^2 distributions are indexed by v, the number of degrees of freedom. In order to examine the shapes of various χ^2 distributions, run program DISTN and select the 'Chi-square distribution' option. Set the number of degrees of freedom to 2. You will see that the pdf is only non-zero for positive values of x (i.e. if the random variable X has the χ_2^2 distribution it cannot take negative values) and the distribution is skewed to the right. These features are common to all χ^2 distributions. Use program DISTN to draw the pdfs of the χ_1^2, χ_3^2, χ_4^2 and χ_5^2 distributions. Note that as the number of degrees of freedom is increased the distribution becomes more spread out and more symmetrical. Clear the screen and plot the pdfs of the χ_{20}^2, χ_{30}^2 and χ_{10}^2 distributions and you will see that the same trend continues for higher degrees of freedom. In fact, for large v, the pdfs of the χ_v^2 and $N(v,2v)$ distributions are similar. For instance, plot the pdf of $N(30,60)$ and check that it is fairly close to the pdf of χ_{30}^2.

Percentage points of χ^2 distributions may be found by using table 4 of Appendix 4. This is laid out in a similar manner to table 3; the columns give p, the rows give v and the figure in the body of the table is the value of x such that $P(X \leqslant x) = p$, where X has the χ_v^2 distribution. As an example, use program DISTN to find the 95% point of χ_{10}^2. (Use the 'Chi-square distribution' and 'Find $100p\%$ point' options.) The computer reports that the 95% point is approximately 18.31. Looking at table 4, for row $v = 10$ and column $p = 0.95$, we see that the 95% point is given as 18.307.

Consider the following example of the occurrence of a contingency table.

Illustration 6.3

Suppose that in a market research study, 290 people were selected at random and asked about their preferences among various types of tablet and other forms of medication. The following table summarises part of the information obtained and relates the ages of the interviewees and their preferred colour of tablets from the range pink, orange and white.

Age group	Colour		
	Pink	Orange	White
18–35	26	40	32
36–60	14	57	49
>60	9	28	35

On the basis of these (artificial) data, can it be concluded that there is a link between age and colour preference in the population as a whole?

We have to decide whether the Null Hypothesis, that the proportions of people preferring the pink, orange and white tablets are the same for the three age groups, is reasonably consistent with the data. If it is not (i.e. there is significant evidence against it) then we can conclude that there does appear to be a link between age and colour preference. In this case, we will need to investigate how the age groups differ.

One approach would be to evaluate the sample proportions in each age group preferring pink, orange and white. For age group 18–35 these are approximately 0.27 (i.e. $26/(26 + 40 + 32)$), 0.41 and 0.33, for age group 36–60 they are 0.12, 0.47 and 0.41 and for the over-60s they are 0.12, 0.39 and 0.49. Although these sets of numbers do differ, this does not automatically mean that the corresponding population proportions differ. As usual, we need to use a statistical test which will indicate whether the evidence of a difference is significant.

We shall investigate the analysis of these data by using program CHISQ. Run this program and select the 'Contingency table' option. We must first set up the table in the computer. Select the 'New data' option and choose the default values for all the labels, numbers of rows and columns and frequencies (19 presses of the RETURN key in all). You should then find the contingency table displayed at the top of the screen, along with an additional row and column each labelled 'Total'.

Select the Evaluate marginal totals' option. You will see that the row and column totals and the grand total are entered into the table. For instance, out of the 290 people interviewed, 125 preferred the orange-coloured tablets.

As in the case of t-tests, we need to know the number of degrees of freedom, i.e. the number of independent pieces of information there are in the table that will help us to decide whether the Null Hypothesis of no link between age and colour preference is acceptable. Select the 'Evaluate degrees of freedom' option. The frequencies in the body of the table are erased and just the totals remain. Notice that although the totals tell us something about the proportions in the sample in each age group and the proportions preferring each colour, they tell us nothing at all about any age/colour preference link. (The row and column totals are said to be *ancillary statistics* with respect to the Null Hypothesis.) Press the space bar and you will see that four of the frequencies are restored. Note that, starting with these four frequencies and the row and column totals, we could reconstruct the whole table of nine frequencies. It follows, therefore, that there are just four independent pieces of information relating to the Null Hypothesis, i.e. there are four degrees of freedom.

Press any key and you will see that the full table is restored. We can now proceed to see whether the Null Hypothesis is acceptable. Rather than comparing the proportions in each age group preferring each of the colours, we shall consider the expected frequencies, assuming that the Null Hypothesis is correct. Select the 'Evaluate expected frequencies' option. The resulting display is reproduced in figure 6.2. You can see that for row 1, column 1 (i.e. age group 18–35, preferring pink), the expected frequency is given as 16.6. The explanation for this is summarised in the lower half of the screen. The proportion of people between 18 and 35 in the sample is 98/290, so that if there were no link between age and colour preference out of the 49 people preferring pink, the expected frequency in the 18–35 age group would be $49 \times 98/290$, i.e. approximately 16.6. (It is usually adequate to evaluate expected frequencies to one decimal place when carrying out this test. We are using the word 'expected' in the technical sense introduced in § 2.3.) Repeated pressing of the space bar reveals how the expected frequencies are evaluated for each of the other eight entries in the table. Go through these slowly until you are confident about the method.

On pressing the space bar following the last expected frequency, you will see the message 'No expected frequencies less than 5. Chi-square tables may be used.' This relates to a condition that must hold if the forthcoming χ^2 test is to provide a good approximation to the true significance. The condition is that the expected values are all greater than 5. ('5' is a 'rule-of-thumb' value.) For our data, the smallest expected value is 12.2, so the condition is satisfied with room to spare.

```
                Colour
              Pink    Orange  White  Total
A 18-35       26        40     32      98
g             16.6
e
  36-60       14        57     49     120
g
r
o >60          9        28     35      72
u
P
  Total      49       125    116     290
```

Row 1 column 1.

Proportion in row 1 = 98/290

Number in column 1 = 49

∴Expected frequency = 49*98/290

= 16.6

Figure 6.2 One of the displays obtained when using the 'Contingency table' option of program CHISQ, with the tablet colour data: an explanation of why the expected frequency of people in the 18–35 age group preferring pink-coloured tablets is 16.6, assuming no association between age group and colour preference.

If the Null Hypothesis of no link between age and colour preference were true, the frequencies that were actually observed (which we shall denote by '*O*' for 'Observed') and the corresponding expected frequencies (denoted by '*E*') would in each case be fairly close, subject to occasional larger discrepancies due to sampling variation. Looking at the table on the screen, we see that in some cases the observed and expected frequencies are close (e.g. age group 36–60 and colour white), whilst in others there are large differences (e.g. age group 18–35 and colour pink). We wish to investigate whether these results are reasonably consistent with the Null Hypothesis.

Select the option 'Significance test' and note that the formula for the test statistic is given as $\Sigma (O - E)^2 / E$. More fully, if O_{ij} is the observed frequency in row i, column j and E_{ij} is the corresponding expected frequency ($i = 1, 2, 3$ and $j = 1, 2, 3$), then the test statistic is

$$\sum_{i=1}^{3} \sum_{j=1}^{3} \frac{(O_{ij} - E_{ij})^2}{E_{ij}}.$$

Note that if the pairs of observed and expected frequencies are all close, suggesting that the Null Hypothesis is correct, the value of this test statistic will be relatively small. On the other hand, any large differences between the observed and expected frequencies will lead to a relatively large value of the test statistic. In short, values of the test statistic much larger than zero tend to suggest that the Null Hypothesis is incorrect.

Press the space bar and you will see how the test statistic is evaluated. Notice that there are nine terms in all in the sum. Press the space bar again and the computations are completed. The terms in the sum have been written out as they will be used in a moment.

We have found that the value of the test statistic is 11.78. It is shown in more advanced texts that if the Null Hypothesis is correct then the distribution of the test statistic is approximately χ_4^2, the 4 being the number of degrees of freedom evaluated above. Since it is only large values of the test statistic that tend to contradict the Null Hypothesis, the significance of the value 11.78 is equal to the area to the right of 11.78 under the χ_4^2 probability density function. On pressing the space bar once more, you will see that the significance is about 0.019, or 1.9%. (Run program DISTN, draw the χ_4^2 distribution and evaluate $P(X > 11.78)$ if you wish to confirm the value 0.019.) Since the significance is less than 0.05, we can conclude that there is significant evidence at the 5% level of an association between age group and colour preference.

We must now investigate the nature of this association. The terms in the sum used in evaluating the test statistic help here; those sums which are large indicate where the association exhibits itself most strongly. Select the 'Evidence of association' option. You will see that the largest term, 5.38, is underlined and the corresponding frequency and expected frequency are 'framed'. Note that the observed frequency greatly exceeds the expected frequency, i.e. there is a greater preference for pink tablets in age group 18–35 than in the population as a whole. Press the space bar: the next largest term, 1.94, is underlined; this corresponds to the 36–60 age group's relative dislike of the pink tablets. Continue pressing the space bar to note that the over-60s have a greater preference for white tablets, whilst these are less favoured by the 18–35 group. The terms that are less than 1 do not indicate any strong departure from the Null Hypothesis and are therefore not indicated.

In summary, we have established that there is evidence at the 5% level of a link between age group and colour preference, primarily as a result of pink being more strongly favoured by the 18–35 age group than by the population as a whole. If you wish to go through the steps of this test again, select the 'Repeat' option.

We now consider what should be done when one or more of the expected frequencies are less than 5. Suppose, for instance, that in Illustration 6.3, many fewer people over 60 had been interviewed, so that the following results had been obtained.

| Age | Colour | | |
group	Pink	Orange	White
18–35	26	40	32
36–60	14	57	49
>60	4	2	7

In order to enter this modified table into program CHISQ, choose the 'New data' option and select the default values for the labels, numbers of rows and columns and all the frequencies except those for row 3. Enter the values 4, 2, 7 for these. Go through the steps as before and notice that one of the expected frequencies (for the over-60s preferring pink) is much less than 5. You will eventually see the message 'Some expected frequencies less than 5. Consider amalgamating classes.' The obvious amalgamation in this case is to combine age groups 36–60 and >60 to give the following table.

| Age | Colour | | |
group	Pink	Orange	White
18–35	26	40	32
>35	18	59	56

Enter this table by using the 'New data' option. You will need to change the number of rows to 2 and the label for row 2, as well as some of the frequencies. You will now find that all the expected frequencies are greater than 5, so that the χ^2 tables may be used. Notice that since we have reduced the number of rows, the number of degrees of freedom has also been reduced. The χ^2 test still indicates significant evidence of an association between age group and colour preference, again primarily due to the enhanced popularity of pink tablets among the 18–35 age group.

Mathematical details

We assume that a sample of n individuals is selected at random from a population and that each sampled individual is then identified as belonging to one of r row categories and one of c column categories. It is found that of the n individuals, O_{ij} belong to row i, column j. The data can thus be summarised in an $r \times c$ table, and the row and column totals included, as shown below.

		Columns			
	1	2	....	c	Total
1	O_{11}	O_{12}	.....	O_{1c}	R_1
2	O_{21}	O_{22}	.....	O_{2c}	R_2
Rows .	.	.	.....	.	.
.	.	.	.....	.	.
r	O_{r1}	O_{r2}	.....	O_{rc}	R_r
Total	C_1	C_2	.....	C_c	n

We further assume that it is desired to test the Null Hypothesis of 'no association' between the row categorisation and the column categorisation (i.e. the population proportions of individuals in each column are the same for each row). Under this hypothesis, the expected frequency in row i, column j, is given by

$$E_{ij} = R_i C_j / n \qquad i = 1, 2, ..., r; \; j = 1, 2, ..., c. \qquad (6.4.1)$$

The test statistic is

$$\chi^2 = \sum_{i=1}^{r} \sum_{j=1}^{c} \frac{(O_{ij} - E_{ij})^2}{E_{ij}}. \qquad (6.4.2)$$

Large values of the test statistic lend support to the Alternative Hypothesis, that the Null Hypothesis is wrong. As long as each E_{ij} is greater than 5, the distribution of the test statistic under the Null Hypothesis is approximately $\chi^2_{(r-1)(c-1)}$. Thus, the significance of the test is approximately equal to $P(X \geqslant \chi^2)$, where X has the $\chi^2_{(r-1)(c-1)}$ distribution. Using the χ^2 table, table 4, it is necessary to check whether χ^2 exceeds the 0.95 point of $\chi^2_{(r-1)(c-1)}$ to establish significance at the 5%

level. If so, go on to check whether x exceeds the 0.99 point and the 0.999 point and draw conclusions in the usual way. If there is significant evidence against the Null Hypothesis, the terms in formula (6.4.2) that are large will indicate where the observed and expected frequencies are particularly divergent. When reporting evidence of association, always comment on how it is exhibited.

When any of the expected frequencies is smaller than 5, the above approximation is no longer satisfactory. It is therefore necessary to amalgamate one or more rows or columns and recalculate the test statistic.

Contingency tables with up to four rows and four columns and frequencies less than 2500 may be analysed by using program CHISQ. After running the program, select the 'Contingency table' option and set up the contingency table by using the 'New data' option. You may then proceed through the test step by step by pressing the RETURN key repeatedly, or you may perform the test immediately by scrolling to the 'Significance test' option and selecting it.

Exercise

The following (artificial) data relate to the eye colour/quality of vision survey of 643 randomly selected people, described at the beginning of this section. The quality of vision is measured on the scale $\{1, 2, 3\}$, '1' indicating perfect vision, '2' indicating slightly impaired vision and '3' indicating poorer levels of vision.

Eye colour	Quality of vision		
	1	2	3
Brown	164	94	50
Blue	105	84	53
Hazel	29	16	8
Grey	20	12	8

Show that, on the basis of these data, there is not significant evidence of a link between quality of vision and eye colour. (Answer first by using formula (6.4.2) and table 4, then check your results by using program CHISQ.)

The situation in the following exercise differs from that previously encountered in as much as separate, independent random samples are taken for each row. It turns out that this does not affect the mechanics of the test, though it is now described as a 'Test of homogeneity' rather than a 'Contingency table test'.

Exercise
Before launching a new product, a detergent manufacturer asks randomly selected volunteers in three regions to try it and report back on how they like it. The summarised results of the survey, showing the frequencies of various responses in the three regions, are given below.

Region	Opinion of detergent			
	Very good	Good	Average	Poor
1	12	34	21	25
2	19	46	25	10
3	6	22	26	16

Is there significant evidence of a difference between the regions in respect of the distribution of opinions about the new product? If so, describe the nature of the difference. (Perform the analysis yourself before running program CHISQ.)

The test statistic is modified in the case of a 2×2 table. (The modification is referred to as 'Yates' correction'.) We shall simplify our notation for this case and assume that the frequency table is as follows.

		Columns		
		1	2	Total
Rows	1	a	b	R_1
	2	c	d	R_2
	Total	C_1	C_2	n

The test statistic is

$$\chi^2 = \frac{n(|ad - bc| - n/2)^2}{R_1 R_2 C_1 C_2}. \qquad (6.4.3)$$

(It can be shown that this is formula (6.4.2) with a continuity correction associated with the approximation of binomial distributions by normal distributions. The notation $|ad - bc|$ denotes the absolute value of

$ad - bc$, i.e. if the value of $ad - bc$ is negative, the negative sign is dropped.) As long as the expected values are each greater than 5, the test statistic will approximately follow the χ_1^2 distribution under the Null Hypothesis, and tend to take larger values under the Alternative. The significance is therefore the area under the χ_1^2 probability density function to the right of χ^2.

If any of the expected frequencies is less than 5, it is necessary to use a different statistical test, namely Fisher's exact test. Details may be found in more advanced texts.

Exercise
In an opinion poll, 125 out of 260 men interviewed said that they would vote for Party X, whilst 118 out of 280 women interviewed said they would vote for that party. Is there significant evidence of a difference in the proportions of men and women in the population intending to vote for Party X? (Set up the frequency table and evaluate test statistic (6.4.3) in order to perform the test, then confirm your results by using program CHISQ.)

A word of warning about this χ^2 test: it can only be applied to frequency data, where the units are randomly selected and each unit contributes exactly once. Thus it cannot be used for a table of measurements, averages or percentages or where individuals might be counted more than once. An example of the last case would occur if interviewees were shown a list of names of insurance companies and asked to say which ones they recognised; some interviewees might recognise none of the names while others might recognise several.

⟩6.5   Goodness-of-fit Test

The goodness-of-fit test is used to examine whether a particular type of probability distribution provides an acceptable fit to some data. The test itself is related to the χ^2 test discussed in the previous section. The way in which it is performed depends partly on whether any parameter has to be estimated. We shall look first at the simpler case when no estimation is required.

Illustration 6.4
The numbers of accidents during three shifts over a period of six months

at a certain factory were as follows:

Shift: 1   2   3   Total
Number of accidents: 20   34   26   80

Each shift is of the same length and involves a similar number of people. Is there significant evidence that the underlying accident rate differs between the shifts?

The Null Hypothesis in this case is that accidents are equally likely to occur in each of the shifts, i.e. if we know that one accident had occurred, then the probability that it occurred in shift i is $\frac{1}{3}$, for $i = 1, 2, 3$. The Alternative Hypothesis is that the probabilities differ in some respect. The question to be resolved is whether the data presented in Illustration 6.4 are reasonably consistent with the Null Hypothesis.

Run program CHISQ and select the 'Goodness-of-fit' option. Although it was neater to make 'Shift' the column classification in the above frequency table, when carrying out the goodness-of-fit test it is more convenient to make it the row classification. Thus, in program CHISQ, set the row classification to 'Shift', the number of rows to 3, the names of the rows to 1, 2 and 3 and the frequencies to 20, 34 and 26. A table of the frequencies is displayed. You are next asked for the 'probability of 1'; this refers to the probability of an accident in shift 1, given that one accident has occurred and that the Null Hypothesis is true. Enter '1/3' and respond similarly to the request for the 'probability of 2'. Since the probabilities must add up to 1, the computer deduces the probability for shift 3.

As you have been entering the Null Hypothesis probabilities, numbers have also been written in two further columns. The column headed 'Exp.' contains the expected frequencies under the Null Hypothesis. These are just equal to the product of the total sample size, 80, and the Null Hypothesis probabilities, $\frac{1}{3}$. In other words, the average number of accidents in each shift over a large number of periods when there had been 80 accidents in all would be about 26.7, assuming that the Null Hypothesis is correct. (Note that, subject to rounding errors, the expected values add up to 80, the sample size.) Each expected value must exceed 5 if the χ^2 test is to be used.

The final column is headed '$(O - E)^2/E$' and indicates the terms in the test statistic formula

$$\sum_{i=1}^{3} \frac{(O_i - E_i)^2}{E_i} \qquad (6.5.1)$$

(cf formula (6.4.2)). Here, O_i is the observed frequency in shift i and E_i is the corresponding expected frequency. Notice that the test statistic will be large if the observed and expected frequencies differ greatly. The value '1.7' which appears in the top row of the final column is calculated from the expression $(20-26.7)^2/26.7$, and the other terms are calculated similarly. Adding them up, we find that the value of the test statistic is 3.7.

Before we can perform the χ^2 test, we need to evaluate the number of degrees of freedom. No parameter was estimated in determining the Null Hypothesis distribution, so enter '0' as the number of estimated parameters. You will see that the number of degrees of freedom is equal to 2, the number of rows less 1. (The number of degrees of freedom is again equal to the number of independent pieces of information throwing light on the Null Hypothesis. If we know the number of accidents in two of the shifts and the total number of accidents, then the number for the other shift can be deduced.)

It can be shown that under the Null Hypothesis the test statistic (6.5.1) has the χ_2^2 distribution, whilst it will tend to be larger under the Alternative Hypothesis. The significance is therefore equal to the probability of a χ_2^2 random variable taking a value greater than 3.7, the value of the test statistic. Press the space bar and you will see that this probability is about 0.157. Since the significance is much higher than 0.05, we conclude that the data are reasonably consistent with the Null Hypothesis of equal accident rates. (Though, as usual, we must not fall into the trap of saying that the Null Hypothesis is 'proved'—patently it is not.)

Now let us consider the rather more complex case when a parameter has to be estimated.

Illustration 6.5
The quality level of a mass-produced item is checked by inspection of randomly selected boxes of five. The results of the inspection of 100 such boxes are summarised below.

Number of defectives per box:	0	1	2	3	4	5
Frequency:	49	18	13	11	7	2

Does the number of defectives per box appear to follow the binomial distribution? (Note: a binomial distribution would result if defectives and non-defectives were randomly intermingled.)

This case differs from the previous one in that the distribution under the Null Hypothesis is not fully specified, i.e. it is $B(5,p)$, but p may take any value. In order to evaluate expected frequencies, however, we need to specify a value for p. We therefore estimate p as described in § 3.2, i.e. by the sample proportion of defectives. In this case, 500 examples of the product have been sampled, of which 115 ($= 0 \times 49 + 1 \times 18 + \ldots + 5 \times 2$) are defective. The estimate of p is therefore $\hat{p} = 115/500 = 0.23$. The estimated binomial probabilities are given by the formula

$$P(X = k) = \binom{5}{k} 0.23^k 0.77^{5-k} \qquad k = 0, 1, \ldots, 5$$

(see § 2.4). Evaluating these probabilities to four decimal places, we obtain

$P(X = 0) = 0.2707$	$P(X = 1) = 0.4042$	$P(X = 2) = 0.2415$
$P(X = 3) = 0.0722$	$P(X = 4) = 0.0108$	$P(X = 5) = 0.0006.$

$$(6.5.3)$$

These probabilities should be approximately equal to the corresponding relative frequencies if the data do come from a binomial distribution. In order to check this, run program FREQ and select 'N' for new data, and '2' for integer data. Set the smallest integer to 0 and the largest to 5 and enter the frequencies as shown above. Select the 'Fit a distribution' option. Press 'B' for binomial, and choose the default values for the parameters, 5 and 0.23. (Note that the default value for p is the observed proportion of defectives.) The binomial probabilities given in (6.5.3) are evaluated and displayed on the barchart. (The y-axis scale is now the relative frequency scale.) We can see at once that there seems to be quite a discrepancy between the relative frequencies and the probabilities.

Let us now see whether there is significant evidence against the data following a binomial distribution by using the goodness-of-fit test. Run program CHISQ and select the 'Goodness-of-fit' option. The row classification name is 'Defectives', the number of rows is 6, the names of the rows are 0, 1, ..., 5 and the frequencies are as given in the above table. Enter the estimated Null Hypothesis probabilities given by (6.5.3).

The expected frequencies are again the product of the sample size, 100, and the estimated Null Hypothesis probabilities. We see that a number of the expected frequencies are less than 5, so that we need to re-enter the data in the following form.

Number of defectives per box:	0	1	2	>2
Frequency:	49	18	13	20

Having done this, re-enter the estimated Null Hypothesis probabilities, $P(X = 0)$, $P(X = 1)$ and $P(X = 2)$, as before. The expected values now all exceed 5 and the test statistic is found to be 51.6.

The number of estimated parameters in this case is 1, since p was estimated from the data. On entering this value you will see that the number of degrees of freedom is equal to 2. This is equal to the number of independent frequencies giving information about the Null Hypothesis, 3 ($= 4 - 1$), less the number of estimated parameters.

Under the Null Hypothesis, therefore, the test statistic will have approximately the χ_2^2 distribution. The significance is the probability of a random variable with this distribution exceeding the observed test statistic value of 51.6. Press the space bar to reveal that the significance is approximately 0.0000, i.e. there is very strong evidence indeed against the Null Hypothesis. On comparing the observed and expected values, we see that this is because there are too many boxes with either 0 defectives or more than 2 defectives. Rather than the defectives being randomly intermingled with the non-defectives, they tend to occur in 'bursts'.

Mathematical details

We assume that a random sample of n observations is available and that it is desired to test the Null Hypothesis that the underlying distribution is of a particular form. The data must first be summarised as frequency data, by splitting up the range into r categories, labelled $1, 2, ..., r$, and counting the frequency of each category. Let O_i be the frequency of the ith category; note that $\Sigma O_i = n$.

It is next necessary to find the probability of each category under the Null Hypothesis. If the distribution is fully specified by the Null Hypothesis this should be straightforward. On the other hand, if the Null Hypothesis merely states the family of the distribution (e.g. binomial or normal) then you must first estimate the parameter(s), as described in Chapter 3. The distribution with estimated parameter(s) is then used to find the estimated probabilities of each category. Let p_i be the (estimated) probability of an observation falling into category i under the Null Hypothesis.

The expected frequency of category i under the Null Hypothesis, E_i, is then equal to np_i. In order that the χ^2 tables can be used later on, it is necessary that each of the expected frequencies exceeds 5. If this is not the case, you must combine two or more categories.

The test statistic is

$$\sum_{i=1}^{r} \frac{(O_i - E_i)^2}{E_i}.$$

This can be seen to be a measure of the 'distance' between the observed and expected values. Under the Null Hypothesis, assuming that the expected values all exceed 5, the test statistic has approximately the χ_v^2 distribution, where the number of degrees of freedom, v, is given by

$$v = r - 1 - \text{number of estimated parameters.}$$

(If $v < 1$ it is not possible to carry out the χ^2 test.) The significance is then the area under the χ_v^2 probability density function to the right of the test statistic value. When using the χ^2 table (table 4 of Appendix 4), it is necessary to compare the test statistic value with the 0.95, 0.99 and 0.999 points of the χ_v^2 distribution in order to decide whether there is significant evidence against the Null Hypothesis at the 5%, 1% and 0.1% levels.

If there is significant evidence that the Null Hypothesis is incorrect, it will often be useful to investigate in what respect the hypothesised distribution fails to provide a good fit. This is achieved by comparing the observed and expected values.

Goodness-of-fit tests with up to eight categories and frequencies each less than 2500 can be carried out by using program CHISQ. After running the program, select the 'Goodness-of-fit' option. You will then be taken through the analysis step by step.

Exercise
According to a genetic theory relating to a particular type of plant, the offspring of plants with pink flowers will have red flowers with probability $\frac{1}{4}$, pink flowers with probability $\frac{1}{2}$ and white flowers with probability $\frac{1}{4}$. In an experiment, of 136 such offspring, 40 have red flowers, 72 have pink flowers and 24 have white flowers. Perform a χ^2 test to decide whether there is significant evidence against the theory. Check your answer by using program CHISQ.

Exercise
Recall the thread data first introduced in § 1.1. In 50 100 m lengths of thread, the following frequencies of imperfections were observed.

Number of imperfections:	0	1	2	3	4	5	6	7	8
Frequency:	6	8	10	12	8	4	0	1	1

Is it reasonable to suppose that the numbers of imperfections follow a Poisson distribution? (Remember to combine categories where necessary in order to ensure that the expected frequencies all exceed 5. Use programs FREQ and CHISQ to check your answer.)

〉 Appendix 1

〉 Running the Programs

〉 BBC Micro and Electron computers with disc drive

Insert the disc in Drive 0, hold down 〈SHIFT〉, then press and release 〈BREAK〉. A menu will be shown. Use the ↑ and ↓ keys to highlight the program that you require and press 〈RETURN〉. The requested program will be loaded (usually starting at address &FOO) and run. (If you prefer, you can use LOAD or CHAIN with the program name.) When you have finished, 〈SHIFT〉/〈BREAK〉 will recall the main menu or 〈BREAK〉 will restore the usual disc system.

If the programs are to be run on a BBC Master which is connected to a network, then the shadow memory should be invoked by *SHADOW.

⟩ Appendix 2

⟩ Use of the Editor and Datafiles

The EDITOR is a program which allows files of data to be created and edited, for use by some of the other programs on the disc. The programs STEM, HIST, SCATTER, TESTS1 and TESTS2 all have the ability to read datafiles. The EDITOR is run, and options selected, in the usual way. To create a file, first select the 'Create a datafile' option. A prompt will then be given for the number of variables (1 or 2) and the name(s) of the variables to be entered. The individual cases may then be entered one by one. When this is complete the data may be edited if mistakes have been made. The 'Add' option allows additional cases to be entered. The 'Delete' option allows the deletion of the observation whose case number is entered. The 'Edit' option allows the values for particular cases to be altered by first entering the case number of the observation to be changed. The listing on the screen is continually updated so that changes are registered immediately. When editing is complete a file is stored simply by selecting the 'Write' option.

If the software is being used in 40-track disc version with the BBC, then there is no free disc space available for the storage of datafiles and a separate disc must be used for this purpose. The data disc must be inserted before the 'Read' or 'Write' options are selected in the EDITOR. Similarly, when reading a datafile from one of the programs, the data disc must be inserted before ⟨RETURN⟩ is pressed at the end of the entered filename. Notice also that the file called LOADER must be copied from the program disc to the data disc.

Four programs (HIST, SCATTER, TESTS1, TESTS2) have simple editing facilities within them. These are based on a list of the observations together with identifying case numbers. In order to change the value of a particular observation, simply enter the corresponding case number, and then the new value. To delete a case, enter the case value preceded by a minus sign, e.g. entering -8 will delete case number 8. To add cases, enter a case number larger than the largest currently displayed, and then enter the value of the new case. If the case number -999 is entered then the entire set of data will be erased in preparation for a completely new set of numbers to be entered. To exit from the editor, simply press 〈RETURN〉 in response to a prompt for a case number.

〉 Appendix 3

〉 Program Descriptions

Very brief descriptions of each of the programs which accompany this book are given below. The following conventions should be borne in mind when using the programs.

(i) Where parameter values are to be entered, a default value is often given in brackets. This default value may be accepted by pressing the 〈RETURN〉 key. If another value is to be used instead. then fractions and other arithmetic expressions may be entered (e.g. 2/3 or SQR(2)).

(ii) Often a choice of options is available. Alternative options may be displayed by pressing the space bar. The options may be cycled until the desired one is displayed. The displayed option can then be selected by pressing the 〈RETURN〉 key.

(iii) If you get 'lost' during the operation of a program, you may restart by pressing 〈ESCAPE〉, followed by R. Parameter values which you have already entered are usually adopted as the new defaults.

〉STEM: construction of a histogram and stem and leaf plots

Data may be read from file or the default data may be used. Files should contain data on only one variable. Any negative values will be ignored.

Options
Histogram
          Animation is used to construct a histogram.
Stem & leaf plot
          Animation is used to construct a stem and leaf plot.

173

Transform the data
> Any transformation which keeps all observations positive
> may be used.

)HIST: histograms and summary statistics

Data may be read from file or the default data may be used. Files should contain data on only one variable. Datasets must contain at least three cases before the program will operate.

Options
Histogram
> A histogram is drawn on the screen.

Change the number of histogram bins
> The number of bins on the horizontal scale is altered. If the
> histogram is on display it is redrawn.

Sample mean and standard deviation
> These two summary statistics are displayed by arrows, or
> removed if already present.

Edit the data
> Cases may be added, deleted or altered. All the data may
> be erased by entering − 999 in response to the case prompt.

Sort the data
> The cases are visually sorted into increasing order.

Sample median and quartiles
> These two summary statistics are displayed by arrows, or
> removed if already present. If the data have not been
> sorted this is done first.

Transform
> The data are transformed according to the entered
> function.

Cumulative frequency plot
> Two versions of the plot are drawn. Percentiles may be
> evaluated.

Display frequencies
> This is available when the histogram has been drawn. Fre-
> quencies are printed on top of each bar.

Normal density
> This is available when the histogram has been drawn. A

scaled normal density is drawn to allow a comparison with the shape of the histogram.

⟩FREQ: frequency data

Two sets of data are available and new data may be entered. Data are either in nominal categories or on an integer scale.

Options

Calculate sample mean and variance (integer data only)
> Graphical and numerical illustrations of the calculations are provided.

Fit a distribution
> A specific distribution may be fitted to the data and compared graphically with the relative frequencies. The uniform distribution is one option. In the case of data whose values are non-negative integers, binomial and Poisson distributions may also be fitted.

⟩B&P: binomial and Poisson probabilities

Options

Binomial distribution
> Binomial distributions with sample size between 1 and 125 may be plotted. Once these have been displayed, Poisson or normal approximations may be superimposed and probabilities listed.

Poisson distribution
> Poisson distributions with means between 0 and 95 may be plotted. Once these have been displayed, the normal approximation may be superimposed and probabilities listed.

⟩DISTN: graphs of probability density functions

The probability density functions (pdfs) of several commonly occurring continuous distributions can be displayed. Up to five pdfs can be drawn on the same graph and the mean and standard deviation of each distribution are evaluated. Probabilities and percentage points associated with distributions can be evaluated and the corresponding areas under the pdfs are indicated.

Options and parameters
Normal distribution

> The mean μ and variance σ^2 (> 0) must be specified. As for the following distributions, if a graph already exists, the pdf is superimposed. If a new graph is required, the 'Clear screen' option should be selected first.

Student's t distribution

> The number of degrees of freedom v ($0 < v \leqslant 1000$) must be specified. For $v = 1$, the mean and variance are not defined and for $v = 2$ the variance is not defined.

Chi-square distribution

> The number of degrees of freedom v ($0 < v \leqslant 1000$) must be specified.

Uniform distribution

> The parameters a and b ($a < b$) must be specified.

Gamma distribution

> The index parameter k (> 0) and the scale parameter v (> 0) must be specified.

Evaluate $P(c < X < d)$

> The values of c and d ($c < d$) must be entered. The value of $P(c < X < d)$ is evaluated, where X has the last displayed distribution, and the appropriate area under the pdf is shaded.

Find $100p\%$ point

> The value of p ($0.0001 < p < 0.9999$) must be entered. The value of c such that $P(X < c) = p$ is evaluated and the appropriate area under the pdf is shaded.

Find a middle $100p\%$ region

> The value of p ($0 < p < 0.9999$) must be entered. The values of c and d such that $P(c < X < d) = p$ and $P(X < c) = (1 - p)/2$ are shown and the appropriate area is shaded.

⟩SAMPLE: simple random sampling

Line segments of randomly varying lengths are simulated and a sample is selected either by the user or at random. The sample mean and standard deviation and the population mean and standard deviation of the lengths are displayed for comparison. The program also generates random numbers and demonstrates the use of random number tables to draw random samples.

Options and parameters
Subjective sampling
> Up to 100 line segments of randomly varying lengths are displayed on the screen. (The lengths are simulated values from the exponential distribution, parameter v, where v is itself simulated from a truncated gamma distribution.) Enter the numbers of the eight line segments which are judged as representative as possible of all the line segments, in respect of their distribution of lengths. The numbers entered must be different and each must correspond to a line segment. After the eighth number is entered, the sample and population means and standard deviations of the lengths are evaluated.

Random sampling
> Line segments are displayed as for the previous option. Eight segments are selected at random and the sample and population means and standard deviations of the lengths are evaluated.

Random number generator
> The population size n ($1 < n < 10^7$) and the sample size k ($0 < k < \min(n, 100)$) must be entered. A random sample of k different numbers from the set $\{1, 2, 3, ..., n\}$ is displayed.

Random number tables
> Six lines of a random number table are displayed. The population size n ($1 < n < 10\,000$) must be entered. The computer then highlights successive parts of the table to illustrate how it is used to select different numbers at random from the set $\{1, 2, 3, ..., n\}$.

$\rangle$CONF: confidence intervals for normal and binomial data

The program demonstrates the evaluation of a confidence interval for normal or binomial data and illustrates how the interval is modified by changes of confidence level or sample size. The program also uses simulation to demonstrate the frequency interpretation of confidence intervals. The robustness of confidence intervals to non-normality can also be investigated.

Options and parameters
Display the data
> The current dataset is displayed. A default dataset (described in Illustration 4.1) is provided.

Change the data
> Indicate by typing N or B whether normal or binomial data are required. If N is typed, enter the number of values in the dataset, n $(0 < n < 65)$, and then enter each value in turn. If B is typed, enter the parameter n (> 0) of the binomial distribution and the observed count X $(0 \leqslant X \leqslant n)$.

Evaluate a confidence interval
> The data are displayed using a line plot. If the data are normally distributed, type Y if the variance is known, or N otherwise. The confidence interval formula is displayed. If the data are normally distributed and the variance is assumed to be known, enter its value. On entering the confidence level $c\%$ $(10 < c < 99.996)$ and pressing the space bar, the confidence interval for the mean (normal data) or the parameter p (binomial data) is evaluated and illustrated above the line plot.

Change the confidence level/sample size
> The confidence level and/or the sample size may be changed and a demonstration similar to the one described in the previous paragraph is provided. If the sample size is changed, the new confidence interval is evaluated assuming the same sample mean and variance (normal data) or the same sample proportion X/n (binomial data) as the data set.

Set up simulation
> If confidence intervals based on the assumption of normally distributed data are to be simulated, enter the sample size n (> 0), type N, enter the mean and variance of the simulated data, respond Y or N depending on whether the

variance is to be assumed known when constructing the confidence intervals, enter the confidence level $c\%$ ($10 < c < 99.996$) and indicate whether the data are to be simulated from a normal, uniform or skewed distribution by typing N, U or S. If confidence intervals based on the assumption of binomially distributed data are to be simulated, enter the value of the parameter n (> 0), type B, enter the value of the parameter p ($0.001 < p < 0.999$) and the confidence level $c\%$ ($10 < c < 99.996$). 40 samples of size n, or 40 observations from the $B(n,p)$ distribution, are simulated within the computer and confidence intervals evaluated using a formula based on the normal or t distributions. The intervals are displayed and a count is kept of the number that include the value of the parameter (μ or p).

Simulate

This option is initially available only following selection of the previous option. The simulation of 40 samples is repeated and the new confidence intervals are displayed. The percentage of all simulated confidence intervals including the parameter (μ or p) since the last use of the previous option is recorded.

⟩NORTEST: statistical test for the value of the population mean

The statistical testing of a specified value of the population mean, based on a random sample of data from a normal distribution with known variance, is demonstrated in various ways. The various quantities associated with the test are the sample size n, the variance σ^2, the value of the mean under the Null Hypothesis, μ_0, and the sample mean $\overline{X}$.

Opening sequence
Value of μ under H_0
          Enter the value of μ_0.
Value of variance σ^2
          Enter the value of σ^2 (> 0.00001).
Form of H_1
          Type 1, 2, 3 or 4 depending on whether H_1 is of the form $\mu \neq \mu_0, \mu > \mu_0, \mu < \mu_0$ or $\mu = \mu_1$. In the last case, the value of μ_1 ($\neq \mu_0$) must be entered.

Sample size
>   Enter the sample size n (>0).

Options and parameters
Display hypothesis test
>   Enter the significance level α $(0.001 < \alpha < 0.5)$. The pdf
>   of $\bar{X}$ under H_0 (i.e. $N(\mu_0, \sigma^2/n)$) is drawn and the critical
>   region of the level-α test is indicated. A flashing message
>   shows what decision is to be made depending on which
>   region $\bar{X}$ belongs to. The flashing messages can be stopped
>   by pressing $\langle$RETURN$\rangle$. Two such diagrams can be
>   displayed on the screen at once by repeated use of this
>   option.

Plot pdf of $\bar{X}$
>   Enter the value of the population mean μ. The pdf of the
>   distribution of $\bar{X}$ (i.e. $N(\mu, \sigma^2/n)$) is drawn.

Single simulation
>   Enter the value of the population mean μ. A value of $\bar{X}$ is
>   simulated and indicated on the x axis of the graph. A
>   message indicates whether H_0 should be rejected at the
>   given significance level on the basis of this value of $\bar{X}$.

Continued simulation
>   Enter the value of the population mean μ. Values of $\bar{X}$ are
>   simulated and indicated on the x axis of the graph. The
>   running percentage of simulated values of $\bar{X}$ falling in the
>   critical region of the test is shown.

Change the sample size
>   Enter a new value of n (> 0). This will be used when the
>   'Display hypothesis test' or 'Evaluate significance for
>   specified $\bar{X}$' options are subsequently chosen.

Change H_1
>   Indicate the new form of H_1 (as described above). This
>   Alternative Hypothesis will be used when the 'Display
>   hypothesis test' or 'Evaluate significance level for specified
> $\bar{X}$' options are subsequently chosen.

Evaluate significance for specified $\bar{X}$.
>   Enter the observed value of $\bar{X}$. The significance of this
>   value is evaluated and its graphical interpretation is
>   displayed.

⟩POWER: the power of a statistical test

This program demonstrates the meaning of the terms 'power' and 'power function' and illustrates how they depend on the choice of Alternative Hypothesis, significance level and sample size. The assumptions about the sample available are as for program NORTEST.

Opening sequence
As for NORTEST

Options and parameters
Value of significance level
> Enter the significance level α $(0.001 < \alpha < 0.5)$ for the test.

Evaluate power at specified value of μ
> The critical region of the test is displayed. Enter a value for the population mean μ. The probability of the test rejecting H_0 (i.e. the power at μ) is evaluated and its graphical representation is displayed.

Change the sample size
> Enter a new value of the sample size n (> 0).

Plot the power
> The power evaluated previously is plotted against the value of μ.

Plot the complete power function
> The power is plotted for all values of μ in the displayed range. Power function plots for different significance levels and sample sizes are superimposed.

⟩BINTEST: statistical test for binomial data

The program demonstrates the construction of a statistical test of a specified value of the parameter p of the $B(n,p)$ distribution and the evaluation of the power of the test.

Opening sequence
Value of p under H_0
> Enter p_0 $(0 < p < 1)$, the value of p under the Null Hypothesis.

Value of n

    Enter the value of the parameter n $(1 < n < 61)$.

Form of H_1

    Type 1, 2, 3 or 4 depending on whether H_1 is of the form $p \neq p_0$, $p > p_0$, $p < p_0$ or $p = p_1$. In the last case the value of p_1 ($\neq p_0$) must be entered.

Options and parameters

Evaluate significance

    The binomial probabilities under H_0 are displayed in histogram form. Enter the observed value of the binomial value X $(0 \leqslant X \leqslant n)$. A graphical demonstration of the evaluation of the significance of this value is given.

Display hypothesis test

    This option is not available for $H_1 : p \neq p_0$. Enter the value of the significance level α $(0.001 < \alpha < 0.3)$. The method of finding the critical region of the test is demonstrated. (Notice that the probability of H_0 being rejected for the resulting test is usually smaller than the specified significance level.)

Evaluate the power

    This option is only available after the previous option has been selected. Enter the value of p $(0 < p < 1)$ at which the power is to be evaluated. The probabilities of the $B(n, p)$ distribution are superimposed on the existing diagram and colour/shading is used to indicate how the power is evaluated.

Normal approximation

    The pdf of the $N(np_0, np_0(1 - p_0))$ distribution is superimposed on the last diagram. (This may be used, for instance, to provide an indication of whether it is legitimate to use a normal approximation in deriving a test.)

Binomial probabilities

    The non-negligible probabilities of the $B(n,p)$ distribution are displayed, where p is initially set to p_0, but may be reset using the 'Evaluate the power' option.

〉TESTS1: hypothesis tests on one-sample or paired-sample data

Data may be read from file or the default data may be used. If a file contains data on two variables then the data are assumed to be paired. For paired data, a line plot is drawn and differences are taken to reduce the data to one sample. Datasets must contain at least three cases before the program will operate.

Options
t-test

A *t*-test is performed, with a chosen Null Hypothesis mean.

Sign test

A sign test is performed and illustrated, with a chosen Null Hypothesis median.

Signed ranks test

A signed ranks test is performed and illustrated, with a chosen Null Hypothesis median.

Edit the data

Cases may be added, deleted or edited. All data may be erased by entering − 999 in response to the case prompt.

〉TESTS2: hypothesis tests on two samples of data

Data may be read from file or the default data may be used. Files should contain data on two variables. Datasets must contain at least three cases in each group before tests can be applied.

Options
Line plot

A line plot of each sample is drawn.

Boxplot

A boxplot of each sample is drawn. Both sample sizes must exceed 5.

Two-sample *t*-test

A two-sample *t*-test of the Null Hypothesis that the two population means are identical is performed.

Mann–Whitney *U*-test

A Mann–Whitney *U*-test of the Null Hypothesis that the two population medians are identical is performed and illustrated by animation.

Edit the data

Cases may be added, deleted or edited. All data may be erased by entering − 999 in response to the case prompt.

⟩SCATTER: scatterplots and simple linear regression

Data may be read from file or the default data may be used. Files should contain data on two variables. A prompt will be given to enable the response variable to be identified. When the data have been read, a scatterplot is drawn.

Options
Regression

> A regression line may be moved about the scatterplot by using the cursor keys. Residuals may be drawn. The least-squares regression line may be drawn and a residual plot constructed by animation. Notice that the graphics for the residuals will be displayed only if all x values are well separated.

Transform x

> The explanatory variable is transformed according to the entered function.

Transform Y

> The response variable is transformed according to the entered function.

Edit the data

> Cases may be added, deleted or edited. All data may be erased by entering -999 in response to the case prompt.

⟩CHISQ: contingency tables and goodness-of-fit tests

The program demonstrates how contingency table and goodness-of-fit tests are carried out. The choice of test is made initially.

Options and parameters for the contingency table test
New data

> Allows the contingency table to be changed. Enter the name of the row classification (up to ten characters), the number of rows (up to four), the labels of the rows (up to six characters each), the name of the column classification (up to ten characters), the number of columns (up to four) and the labels of the columns (up to six characters each). Next enter the frequencies into the table (between 0 and 2499 inclusive).

Evaluate marginal totals

> Only available after the 'New data' and 'Repeat' options.
> Adds the row and column totals to the contingency table.

Evaluate degrees of freedom

> Only available after the previous option. Indicates the interpretation of the degrees of freedom as the number of independent pieces of information in the table, given the marginal totals.

Evaluate expected frequencies

> Only available after the previous option. Shows how the expected frequencies, assuming no association between rows and columns, are evaluated. A message indicates whether any expected frequency is less than 5; if so, the chi-square tables may provide a poor approximation to the significance.

Significance test

> The calculations required to evaluate the test statistic are demonstrated and the approximate significance is given. If the table has two rows and columns, Yates' correction is used.

Evidence of association

> Only available after the previous option. Highlights the entries in the two-way table where the observed and expected frequencies differ, in decreasing order of contribution to the test statistic. Those entries whose contribution to the test statistic is less than 1 are not highlighted.

Options and parameters for the goodness-of-fit test

Name of row classification

> Enter a title for the data (up to ten characters)

Number of rows

> Enter the number of classes (up to eight).

Name of row

> Enter the label of the class (up to six characters).

Frequency of row

> Enter the frequency of the class (between 0 and 2499 inclusive).

Probability

> Enter the Null Hypothesis probability of each class, except for the last class which is deduced by the computer. Prob-

abilities must lie between 0 and 1, inclusive, and add to 1.
On entering each probability, the expected frequency
under the Null Hypothesis and the contribution to the test
statistic are evaluated. After the last probability is entered,
the test statistic is evaluated and a message indicates
whether any expected frequency is less than 5. (If so, the
chi-square tables may provide a poor approximation to the
significance.)

Number of estimated parameters

Enter the number of parameters which had to be estimated
from the data when fitting the Null Hypothesis distribution
in order to evaluate the probabilities.

Evaluate significance

The approximate significance of the test statistic is
evaluated from chi-square tables.

〉EDITOR: creating and editing datafiles

Files of data may be created for storage on disc and later use by other
programs. Files which have already been stored may be modified.

Options

Create a datafile

Prompts are given for the number (1 or 2) and the names
of variables on which information is to be entered. The
values of individual observations may then be entered.
When this is complete, scrolling of options is restarted by
pressing 〈RETURN〉. A maximum of 64 cases may be
entered with one variable, and a maximum of 25 cases with
two variables.

Read from a datafile

When the name of a file is entered, the stored data will be
read (if a file of this name exists) and displayed on the
screen.

Edit

If the case number of an observation is entered then the
opportunity is given to alter the values of this case.

Add new cases

The values of new observations may be entered. Scrolling
of options may be restarted by pressing 〈RETURN〉.

Delete cases

> The observations whose case numbers are entered are deleted from the file. Scrolling of options may be restarted by pressing ⟨RETURN⟩.

Change variable names

> New names may be entered.

Write to a datafile

> The current data will be stored in the named file. If a file of this name already exists, the opportunity will be given to use another name or to overwrite the file.

〉 Appendix 4

〉 Statistical Tables

Table 1 Standard normal: distribution function ($\Phi(z)$).

z	$\Phi(z)$									
	0	1	2	3	4	5	6	7	8	9
0.0	0.5000	5040	5080	5120	5160	5199	5239	5279	5319	5359
0.1	0.5398	5438	5478	5517	5557	5596	5636	5675	5714	5753
0.2	0.5793	5832	5871	5910	5948	5987	6026	6064	6103	6141
0.3	0.6179	6217	6255	6293	6331	6368	6406	6443	6480	6517
0.4	0.6554	6591	6628	6664	6700	6736	6772	6808	6844	6879
0.5	0.6915	6950	6985	7019	7054	7088	7123	7157	7190	7224
0.6	0.7257	7291	7324	7357	7389	7422	7454	7486	7517	7549
0.7	0.7580	7611	7642	7673	7704	7734	7764	7794	7823	7852
0.8	0.7881	7910	7939	7967	7995	8023	8051	8078	8106	8133
0.9	0.8159	8186	8212	8238	8264	8289	8315	8340	8365	8389
1.0	0.8413	8438	8461	8485	8508	8531	8554	8577	8599	8621
1.1	0.8643	8665	8686	8708	8729	8749	8770	8790	8810	8830
1.2	0.8849	8869	8888	8907	8925	8944	8962	8980	8997	9015
1.3	0.9032	9049	9066	9082	9099	9115	9131	9147	9162	9177
1.4	0.9192	9207	9222	9236	9251	9265	9279	9292	9306	9319
1.5	0.9332	9345	9357	9370	9382	9394	9406	9418	9429	9441
1.6	0.9452	9463	9474	9484	9495	9505	9515	9525	9535	9545
1.7	0.9554	9564	9573	9582	9591	9599	9608	9616	9625	9633
1.8	0.9641	9649	9656	9664	9671	9678	9686	9693	9699	9706
1.9	0.9713	9719	9726	9732	9738	9744	9750	9756	9761	9767
2.0	0.9772	9778	9783	9788	9793	9798	9803	9808	9812	9817
2.1	0.9821	9826	9830	9834	9838	9842	9846	9850	9854	9857
2.2	0.9861	9864	9868	9871	9875	9878	9881	9884	9887	9890
2.3	0.9893	9896	9898	9901	9904	9906	9909	9911	9913	9916
2.4	0.9918	9920	9922	9925	9927	9929	9931	9932	9934	9936
2.5	0.9938	9940	9941	9943	9945	9946	9948	9949	9951	9952
2.6	0.9953	9955	9956	9957	9959	9960	9961	9962	9963	9964
2.7	0.9965	9966	9967	9968	9969	9970	9971	9972	9973	9974
2.8	0.9974	9975	9976	9977	9977	9978	9979	9979	9980	9981
2.9	0.9981	9982	9982	9983	9984	9984	9985	9985	9986	9986
3.0	0.9987	9987	9987	9988	9988	9989	9989	9989	9990	9990
3.1	0.9990	9991	9991	9991	9992	9992	9992	9992	9993	9993
3.2	0.9993	9993	9994	9994	9994	9994	9994	9995	9995	9995
3.3	0.9995	9995	9995	9996	9996	9996	9996	9996	9996	9997
3.4	0.9997	9997	9997	9997	9997	9997	9997	9997	9997	9998
3.5	0.9998	9998	9998	9998	9998	9998	9998	9998	9998	9998

Table 2 Standard normal: percentage points (z_p).

p	z_p	p	z_p	p	z_p
0.50	0.0000	0.966	1.8250	0.986	2.1973
0.60	0.2533	0.968	1.8522	0.987	2.2262
0.70	0.5244	0.970	1.8808	0.988	2.2571
0.80	0.8416	0.971	1.8957	0.989	2.2904
0.85	1.0364	0.972	1.9110	0.990	2.3263
0.90	1.2816	0.973	1.9268	0.991	2.3656
0.91	1.3408	0.974	1.9431	0.992	2.4089
0.92	1.4051	0.975	1.9600	0.993	2.4573
0.93	1.4758	0.976	1.9774	0.994	2.5121
0.94	1.5548	0.977	1.9954	0.995	2.5758
0.950	1.6449	0.978	2.0141	0.996	2.6521
0.952	1.6646	0.979	2.0335	0.997	2.7478
0.954	1.6849	0.980	2.0537	0.998	2.8782
0.956	1.7060	0.981	2.0749	0.999	3.0902
0.958	1.7279	0.982	2.0969	0.999 50	3.2905
0.960	1.7507	0.983	2.1201	0.999 90	3.7190
0.962	1.7744	0.984	2.1444	0.999 95	3.8906
0.964	1.7991	0.985	2.1701	0.999 99	4.2649

Table 3 Student's t: percentage points ($t_{v,p}$).

v			p			
	0.95	0.975	0.99	0.995	0.999	0.9995
1	6.3138	12.7062	31.8205	63.656	318.308	636.6192
2	2.9200	4.3027	6.9646	9.924	22.327	31.5991
3	2.3534	3.1824	4.5407	5.8409	10.2145	12.9240
4	2.1318	2.7763	3.7470	4.6041	7.1732	8.6103
5	2.0150	2.5706	3.3648	4.0322	5.8934	6.8688
6	1.9432	2.4469	3.1426	3.7074	5.2076	5.9588
7	1.8946	2.3646	2.9979	3.4995	4.7851	5.4079
8	1.8595	2.3060	2.8965	3.3554	4.5007	5.0412
9	1.8331	2.2622	2.8214	3.2498	4.2968	4.7809
10	1.8125	2.2281	2.7638	3.1693	4.1437	4.5869
11	1.7959	2.2010	2.7181	3.1058	4.0247	4.4370
12	1.7823	2.1788	2.6810	3.0545	3.9296	4.3178
13	1.7709	2.1604	2.6503	3.0123	3.8520	4.2208
14	1.7613	2.1448	2.6245	2.9768	3.7874	4.1405
15	1.7531	2.1314	2.6025	2.9467	3.7328	4.0728
16	1.7459	2.1199	2.5835	2.9208	3.6862	4.0150
17	1.7396	2.1098	2.5669	2.8982	3.6458	3.9651
18	1.7341	2.1009	2.5524	2.8784	3.6105	3.9216

Continued

Table 3 (continued)

v	\multicolumn{6}{c	}{p}
	0.95	0.975	0.99	0.995	0.999	0.9995
19	1.7291	2.0930	2.5395	2.8609	3.5794	3.8834
20	1.7247	2.0860	2.5280	2.8453	3.5518	3.8495
21	1.7207	2.0796	2.5176	2.8314	3.5272	3.8193
22	1.7171	2.0739	2.5083	2.8188	3.5050	3.7921
23	1.7139	2.0687	2.4999	2.8073	3.4850	3.7676
24	1.7109	2.0639	2.4922	2.7969	3.4668	3.7454
25	1.7081	2.0595	2.4851	2.7874	3.4502	3.7251
26	1.7056	2.0555	2.4786	2.7787	3.4350	3.7066
27	1.7033	2.0518	2.4727	2.7707	3.4210	3.6896
28	1.7011	2.0484	2.4671	2.7633	3.4082	3.6739
29	1.6991	2.0452	2.4620	2.7564	3.3962	3.6594
30	1.6973	2.0423	2.4573	2.7500	3.3852	3.6460
31	1.6955	2.0395	2.4528	2.7440	3.3749	3.6335
32	1.6939	2.0369	2.4487	2.7385	3.3653	3.6218
33	1.6924	2.0345	2.4448	2.7333	3.3563	3.6109
34	1.6909	2.0322	2.4411	2.7284	3.3479	3.6007
35	1.6896	2.0301	2.4377	2.7238	3.3400	3.5911
36	1.6883	2.0281	2.4345	2.7195	3.3326	3.5821
37	1.6871	2.0262	2.4314	2.7154	3.3256	3.5737
38	1.6860	2.0244	2.4286	2.7116	3.3190	3.5657
39	1.6849	2.0227	2.4258	2.7079	3.3128	3.5581
40	1.6839	2.0211	2.4233	2.7045	3.3069	3.5510
42	1.6820	2.0181	2.4185	2.6981	3.2960	3.5377
44	1.6802	2.0154	2.4141	2.6923	3.2861	3.5258
46	1.6787	2.0129	2.4102	2.6870	3.2771	3.5150
48	1.6772	2.0106	2.4066	2.6822	3.2689	3.5051
50	1.6759	2.0086	2.4033	2.6778	3.2614	3.4960
55	1.6730	2.0040	2.3961	2.6682	3.2451	3.4764
60	1.6706	2.0003	2.3901	2.6603	3.2317	3.4602
65	1.6686	1.9971	2.3851	2.6536	3.2204	3.4466
70	1.6669	1.9944	2.3808	2.6479	3.2108	3.4350
75	1.6654	1.9921	2.3771	2.6430	3.2025	3.4250
80	1.6641	1.9901	2.3739	2.6387	3.1953	3.4163
85	1.6630	1.9883	2.3710	2.6349	3.1889	3.4087
90	1.6620	1.9867	2.3685	2.6316	3.1833	3.4019
95	1.6611	1.9853	2.3662	2.6286	3.1782	3.3959
100	1.6602	1.9840	2.3642	2.6259	3.1737	3.3905
125	1.6571	1.9791	2.3565	2.6157	3.1567	3.3701
150	1.6551	1.9759	2.3515	2.6090	3.1455	3.3566
175	1.6536	1.9736	2.3478	2.6042	3.1375	3.3470
200	1.6525	1.9719	2.3451	2.6006	3.1315	3.3398
400	1.6487	1.9659	2.3357	2.5882	3.1107	3.3150

Table 4 Chi-squared distribution: percentage points ($\chi^2_{v,p}$).

v	\multicolumn{11}{c}{p}
	0.005	0.01	0.025	0.05	0.10	0.50	0.90	0.95	0.975	0.99	0.995	0.999
1	.000	.000	.001	.004	.016	.455	2.706	3.841	5.024	6.635	7.879	10.828
2	.010	.020	.051	.103	.211	1.386	4.605	5.991	7.378	9.210	10.597	13.816
3	.072	.115	.216	.352	.584	2.366	6.251	7.815	9.348	11.345	12.838	16.266
4	.207	.297	.484	.711	1.064	3.357	7.779	9.488	11.143	13.277	14.860	18.467
5	.412	.554	.831	1.145	1.610	4.351	9.236	11.070	12.833	15.086	16.750	20.515
6	.676	.872	1.237	1.635	2.204	5.348	10.645	12.592	14.449	16.812	18.548	22.458
7	.989	1.239	1.690	2.167	2.833	6.346	12.017	14.067	16.013	18.475	20.278	24.322
8	1.344	1.646	2.180	2.733	3.490	7.344	13.362	15.507	17.535	20.090	21.955	26.124
9	1.735	2.088	2.700	3.325	4.168	8.343	14.684	16.919	19.023	21.666	23.589	27.877
10	2.156	2.558	3.247	3.940	4.865	9.342	15.987	18.307	20.483	23.209	25.188	29.588
11	2.603	3.053	3.816	4.575	5.578	10.341	17.275	19.675	21.920	24.725	26.757	31.264
12	3.074	3.571	4.404	5.226	6.304	11.340	18.549	21.026	23.337	26.217	28.300	32.909
13	3.565	4.107	5.009	5.892	7.042	12.340	19.812	22.362	24.736	27.688	29.819	34.528
14	4.075	4.660	5.629	6.571	7.790	13.339	21.064	23.685	26.119	29.141	31.319	36.123
15	4.601	5.229	6.262	7.261	8.547	14.339	22.307	24.996	27.488	30.578	32.801	37.697
16	5.142	5.812	6.908	7.962	9.312	15.338	23.542	26.296	28.845	32.000	34.267	39.252
17	5.697	6.408	7.564	8.672	10.085	16.338	24.769	27.587	30.191	33.409	35.718	40.790
18	6.265	7.015	8.231	9.390	10.865	17.338	25.989	28.869	31.526	34.805	37.156	42.312
19	6.844	7.633	8.907	10.117	11.651	18.338	27.204	30.144	32.852	36.191	38.582	43.820
20	7.434	8.260	9.591	10.851	12.443	19.337	28.412	31.410	34.170	37.566	39.997	45.315
21	8.034	8.897	10.283	11.591	13.240	20.337	29.615	32.671	35.479	38.932	41.401	46.797
22	8.643	9.542	10.982	12.338	14.041	21.337	30.813	33.924	36.781	40.289	42.796	48.268
23	9.260	10.196	11.689	13.091	14.848	22.337	32.007	35.172	38.076	41.638	44.181	49.728
24	9.886	10.856	12.401	13.848	15.659	23.337	33.196	36.415	39.364	42.980	45.559	51.179
25	10.520	11.524	13.120	14.611	16.473	24.337	34.382	37.652	40.646	44.314	46.928	52.620

26	11.160	12.198	13.844	15.379	17.292	25.336	35.563	38.885	41.923	45.642	48.290	54.052
27	11.808	12.879	14.573	16.151	18.114	26.336	36.741	40.113	43.195	46.963	49.645	55.476
28	12.461	13.565	15.308	16.928	18.939	27.336	37.916	41.337	44.461	48.278	50.993	56.892
29	13.121	14.256	16.047	17.708	19.768	28.336	39.087	42.557	45.722	49.588	52.336	58.301
30	13.787	14.953	16.791	18.493	20.599	29.336	40.256	43.773	46.979	50.892	53.672	59.703
31	14.458	15.655	17.539	19.281	21.434	30.336	41.422	44.985	48.232	52.191	55.003	61.098
32	15.134	16.362	18.291	20.072	22.271	31.336	42.585	46.194	49.480	53.486	56.328	62.487
33	15.815	17.074	19.047	20.867	23.110	32.336	43.745	47.400	50.725	54.776	57.648	63.870
34	16.501	17.789	19.806	21.664	23.952	33.336	44.903	48.602	51.966	56.061	58.964	65.247
35	17.192	18.509	20.569	22.465	24.797	34.336	46.059	49.802	53.203	57.342	60.275	66.619
36	17.887	19.233	21.336	23.269	25.643	35.336	47.212	50.998	54.437	58.619	61.581	67.985
37	18.586	19.960	22.106	24.075	26.492	36.336	48.363	52.192	55.668	59.893	62.883	69.346
38	19.289	20.691	22.878	24.884	27.343	37.335	49.513	53.384	56.896	61.162	64.181	70.703
39	19.996	21.426	23.654	25.695	28.196	38.335	50.660	54.572	58.120	62.428	65.476	72.055
40	20.707	22.164	24.433	26.509	29.051	39.335	51.805	55.758	59.342	63.691	66.766	73.402
45	24.311	25.901	28.366	30.612	33.350	44.335	57.505	61.656	65.410	69.957	73.166	80.077
50	27.991	29.707	32.357	34.764	37.689	49.335	63.167	67.505	71.420	76.154	79.490	86.661
60	35.534	37.485	40.482	43.188	46.459	59.335	74.397	79.082	83.298	88.379	91.952	99.607
70	43.275	45.442	48.758	51.739	55.329	69.334	85.527	90.531	95.023	100.425	104.215	112.317
80	51.172	53.540	57.153	60.391	64.278	79.334	96.578	101.879	106.629	112.329	116.321	124.839
90	59.196	61.754	65.647	69.126	73.291	89.334	105.565	113.145	118.136	124.116	128.299	137.208
100	67.328	70.065	74.222	77.929	82.358	99.334	118.498	124.342	129.561	135.807	140.169	149.449
110	75.550	78.458	82.867	86.792	91.471	109.334	129.385	135.480	140.917	147.414	151.948	161.581
120	83.829	86.909	91.568	95.705	100.627	119.335	140.228	146.565	152.214	158.962	163.670	173.668
130	92.201	95.437	100.326	104.662	109.814	129.334	151.041	157.608	163.456	170.435	175.299	185.619
140	100.634	104.021	109.132	113.659	119.032	139.334	161.823	168.611	174.650	181.852	186.867	197.497
150	109.122	112.655	117.980	122.692	128.278	149.334	172.577	179.579	185.803	193.219	198.380	209.310
160	117.660	121.333	126.866	131.756	137.549	159.334	183.307	190.515	196.918	204.541	209.843	221.062
170	126.243	130.053	135.786	140.849	146.842	169.334	194.014	201.422	207.998	215.822	221.261	232.762
180	134.866	138.809	144.737	149.969	156.156	179.334	204.700	212.302	219.047	227.066	232.638	244.411
190	143.528	147.599	153.717	159.113	165.488	189.334	215.367	223.159	230.067	238.276	243.977	256.016
200	152.224	156.421	162.724	168.279	174.838	199.334	226.017	233.993	241.060	249.455	255.281	267.579

Table 5 Critical regions for the sign test. Null Hypothesis, $H_0: m = m_0$, where m denotes the median of the distribution. Test statistic = number of observations which are greater than m_0. These critical regions were calculated by using program BINTEST.

Sample size n	Significance level 5%			Significance level 1%		
	$H_1: m \neq m_0$	$H_1: m < m_0$	$H_1: m > m_0$	$H_1: m \neq m_0$	$H_1: m < m_0$	$H_1: m > m_0$
5	—	0	5	—	—	—
6	0 or 6	0	6	—	—	—
7	0 or 7	0	7	—	0	7
8	0 or 8	$\leqslant 1$	$\geqslant 7$	0 or 8	0	8
9	$\leqslant 1$ or $\geqslant 8$	$\leqslant 1$	$\geqslant 8$	0 or 9	0	9
10	$\leqslant 1$ or $\geqslant 9$	$\leqslant 1$	$\geqslant 9$	0 or 10	0	10
11	$\leqslant 1$ or $\geqslant 10$	$\leqslant 2$	$\geqslant 9$	0 or 11	$\leqslant 1$	$\geqslant 10$
12	$\leqslant 2$ or $\geqslant 10$	$\leqslant 2$	$\geqslant 10$	$\leqslant 1$ or $\geqslant 11$	$\leqslant 1$	$\geqslant 11$
13	$\leqslant 2$ or $\geqslant 11$	$\leqslant 3$	$\geqslant 10$	$\leqslant 1$ or $\geqslant 12$	$\leqslant 1$	$\geqslant 12$
14	$\leqslant 2$ or $\geqslant 12$	$\leqslant 3$	$\geqslant 11$	$\leqslant 1$ or $\geqslant 13$	$\leqslant 2$	$\geqslant 12$
15	$\leqslant 3$ or $\geqslant 12$	$\leqslant 3$	$\geqslant 12$	$\leqslant 2$ or $\geqslant 13$	$\leqslant 2$	$\geqslant 13$

Table 6 Critical regions for the Wilcoxon signed ranks test. Null Hypothesis, $H_0: m = m_0$, where m denotes the median of the distribution. The distribution is assumed to be symmetric. Test statistic = sum of the ranks of the *positive* differences, $X_i - m_0$. These critical regions were calculated from the formulae given in an article by F Wilcoxon (1945 Individual comparisons by ranking methods *Biometrics* 1 80–3).

Sample size n	Significance level 5%			Significance level 1%		
	$H_1: m \neq m_0$	$H_1: m < m_0$	$H_1: m > m_0$	$H_1: m \neq m_0$	$H_1: m < m_0$	$H_1: m > m_0$
5	—	0	15	—	—	—
6	0 or ≥ 21	≤ 2	≥ 19	—	—	—
7	≤ 2 or ≥ 26	≤ 3	≥ 25	0 or 36	0	28
8	≤ 3 or ≥ 33	≤ 5	≥ 31	≤ 1 or ≥ 44	≤ 1	≥ 35
9	≤ 5 or ≥ 40	≤ 8	≥ 37	≤ 3 or ≥ 52	≤ 3	≥ 42
10	≤ 8 or ≥ 47	≤ 10	≥ 45	≤ 5 or ≥ 61	≤ 5	≥ 50
11	≤ 10 or ≥ 56	≤ 13	≥ 53	≤ 7 or ≥ 71	≤ 7	≥ 59
12	≤ 13 or ≥ 65	≤ 17	≥ 61	≤ 9 or ≥ 82	≤ 9	≥ 69
13	≤ 17 or ≥ 74	≤ 21	≥ 70	≤ 12 or ≥ 93	≤ 12	≥ 79
14	≤ 21 or ≥ 84	≤ 25	≥ 80	≤ 15 or ≥ 105	≤ 15	≥ 90
15	≤ 25 or ≥ 95	≤ 30	≥ 90		≤ 19	≥ 101

Table 7 Critical regions for the Mann–Whitney U-test. Null Hypothesis, $H_0 : m_x = m_y$, where m_x and m_y denote the medians of the two distributions. The two distributions are assumed to have the same shape. Alternative Hypothesis, $H_1 : m_x \neq m_y$. n_x and n_y denote the sizes of each sample. Test statistic = smaller of U_x and U_y, where $U_x = n_x n_y + n_x \times (n_x + 1)/2 -$ (sum of the ranks of the x's), and $U_y = n_x n_y + n_y(n_y + 1)/2 -$ (sum of the ranks of the y's). For each n_x and n_y, the critical region is given by the values less than or equal to the tabulated value. These critical regions were calculated from the formulae given in an article by H B Mann and D R Whitney (1947 On a test of whether one of two variables is stochastically larger than the other *Annals of Mathematical Statistics* **18** 50–60).

Significance level 5%

n_x	$n_y = 3$	4	5	6	7	8	9	10	11	12	13	14	15
3	—	—	0	1	1	2	2	3	3	4	4	5	5
4	—	0	1	2	3	4	4	5	6	7	8	9	10
5	0	1	2	3	5	6	7	8	9	11	12	13	14
6	1	2	3	5	6	8	10	11	13	14	16	17	19
7	1	3	5	6	8	10	12	14	16	18	20	22	24
8	2	4	6	8	10	13	15	17	19	22	24	26	29
9	2	4	7	10	12	15	17	20	23	26	28	31	34
10	3	5	8	11	14	17	20	23	26	29	33	36	39
11	3	6	9	13	16	19	23	26	30	33	37	40	44
12	4	7	11	14	18	22	26	29	33	37	41	45	49
13	4	8	12	16	20	24	28	33	37	41	45	50	54
14	5	9	13	17	22	26	31	36	40	45	50	55	59
15	5	10	14	19	24	29	34	39	44	49	54	59	64

Significance level 1%

n_x	$n_y = 3$	4	5	6	7	8	9	10	11	12	13	14	15
3	—	—	—	—	—	—	0	0	0	1	1	1	2
4	—	—	—	0	0	1	1	2	2	3	3	4	5
5	—	—	0	1	1	2	3	4	5	6	7	7	8
6	—	0	1	2	3	4	5	6	7	9	10	11	12
7	—	0	1	3	4	6	7	9	10	12	13	15	16
8	—	1	2	4	6	7	9	11	13	15	17	18	20
9	0	1	3	5	7	9	11	13	16	18	20	22	24
10	0	2	4	6	9	11	13	16	18	21	24	26	29
11	0	2	5	7	10	13	16	18	21	24	27	30	33
12	1	3	6	9	12	15	18	21	24	27	31	34	37
13	1	3	7	10	13	17	20	24	27	31	34	38	42
14	1	4	7	11	15	18	22	26	30	34	38	42	46
15	2	5	8	12	16	20	24	29	33	37	42	46	51

Table 8 Random numbers.

33631	75029	30078	06558	24366
32489	12895	58623	22576	29683
68078	94200	25488	50227	25296
47638	93121	57877	05372	03488
80645	81474	67734	11421	07340
55271	80354	73960	44187	28292
73457	81905	17208	33334	27496
79212	03075	18562	78970	99921
20393	41247	59773	92437	76126
59253	47653	66119	97684	21440
09329	79760	72868	56297	17561
12861	62861	99832	63543	95005
58359	43944	29683	12293	25115
91558	29125	27852	79334	39768
40300	46340	14706	47293	75479
29119	88345	27257	99355	59633
62081	20762	88543	95239	97729
15228	45884	93119	30992	68483
29228	14704	42167	42020	47838
85966	09483	20804	67609	90286
55612	40282	93956	66278	11234
22652	59411	06068	39061	78019
61521	17496	74328	78213	65439
24324	73769	39116	84865	43557
94814	16817	52637	77382	84490
91336	07167	09099	02731	34563
33348	41108	20136	37156	45746
84236	85409	06987	45415	91604
90843	68327	28890	51307	09319
79978	98600	34260	48632	00927
78408	24749	50478	70807	54442
62949	43702	00018	20691	85079
52052	64179	74454	35109	53873
25034	66602	30475	78242	30408
56453	57250	65386	80339	90925
77070	75731	65081	85950	94500
68371	36546	82935	31775	60436
61559	23973	05681	04979	34087
99433	49220	58196	67382	40322
63804	52803	89066	32388	56378
78543	66129	53138	36242	72121
74963	89850	87496	94762	92099
85822	99419	18636	07012	74372
22278	01351	79288	73669	60584

⟩ Bibliography

The books listed below provide additional mathematical background, applications and exercises on the material covered in the text.

Francis A 1979 *Advanced Level Statistics* (Cheltenham: Stanley Thornes)

Hogg R V H and Tanis E A 1983 *Probability and Statistical Inference* 2nd edn (New York: Macmillan)

Mood A M, Graybill F A and Boes D C 1974 *Introduction to the Theory of Statistics* 3rd edn (Kogakusha, Tokyo: McGraw-Hill)

Robinson D R and Bowman A W 1986 *Introduction to Probability: a Computer Illustrated Text* (Bristol: Adam Hilger)

Wetherill G B 1982 *Elementary Statistical Methods* (London: Chapman and Hall)

The following books discuss in detail the graphical exploration and presentation of data.

Chapman M 1986 *Plain Figures* (London: HMSO)

Huff D *How to Lie with Statistics* (London: Penguin)

Velleman P F and Hoaglin D C 1981 *Applications, Basics, and Computing of Exploratory Data Analysis* (Boston: Duxbury)

Finally, on the topics of statistical computing, simulation, and approximation of functions, the following books are very useful.

Abramowitz M and Stegun I A 1984 *Pocketbook of Mathematical Functions* (Frankfurt am Main: Verlag Harri Deutsch)

Cooke D, Craven A H and Clarke G M 1982 *Basic Statistical Computing* (London: Edward Arnold)

Kennedy W J and Gentle J E 1980 *Statistical Computing* (New York: Marcel Dekker)
Morgan B J T 1984 *Elements of Simulation* (London: Chapman and Hall)

〉 Index